W0042247

The Cambridge Manuals of Science and
Literature

THE WORK OF RAIN AND RIVERS

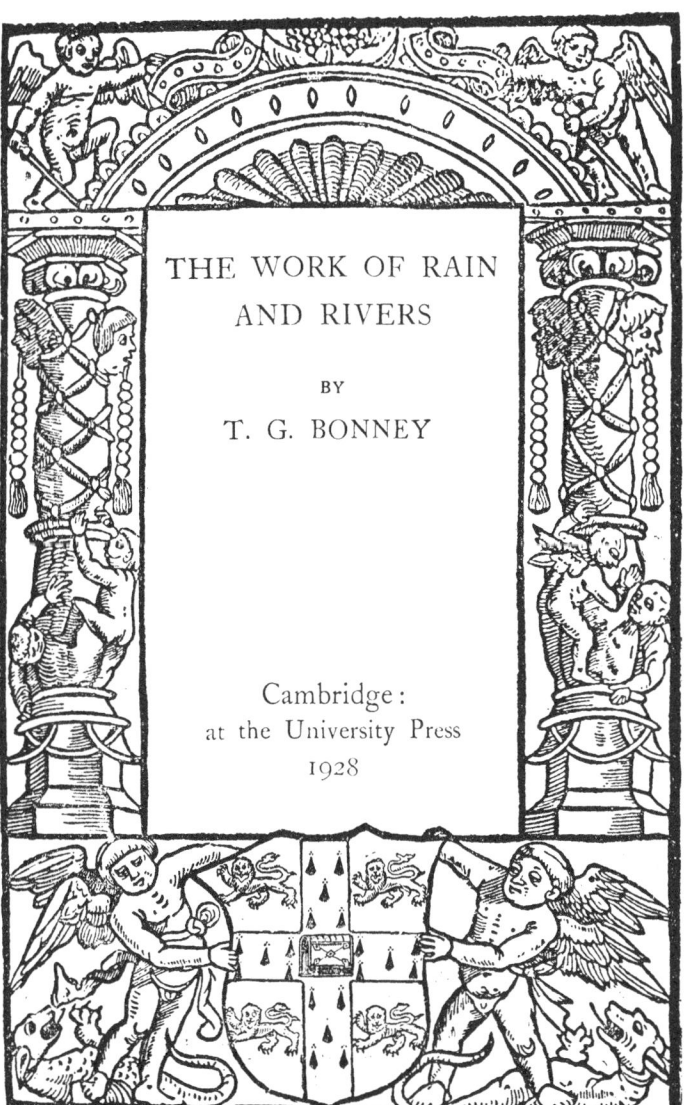

THE WORK OF RAIN AND RIVERS

BY

T. G. BONNEY

Cambridge:
at the University Press
1928

CAMBRIDGE UNIVERSITY PRESS
Cambridge, New York, Melbourne, Madrid, Cape Town,
Singapore, São Paulo, Delhi, Tokyo, Mexico City

Cambridge University Press
The Edinburgh Building, Cambridge CB2 8RU, UK

Published in the United States of America by
Cambridge University Press, New York

www.cambridge.org
Information on this title: www.cambridge.org/9781107401815

© Cambridge University Press 1912

First edition 1912
Reprinted 1928
First paperback edition 2011

A catalogue record for this publication is available from the British Library

ISBN 978-1-107-40181-5 Paperback

Cambridge University Press has no responsibility for the persistence or
accuracy of URLs for external or third-party internet websites referred to in
this publication, and does not guarantee that any content on such websites is,
or will remain, accurate or appropriate.

*With the exception of the coat of arms
at the foot, the design on the title page is a
reproduction of one used by the earliest known
Cambridge printer, John Siberch, 1521*

PREFACE

SO much has been written during the last forty years on the work of rain and rivers, that the author of a brief account of this subject cannot hope to give any new information. But he may do a real service to persons of ordinary education if he can indicate, in language intelligible to them, the methods of observing, the processes of reasoning, and the conclusions reached by trained students of science; for the leading principles in this branch of Geology are not difficult to understand, and the application of them to the interpretation of scenery is an endless source of pleasure. Thus in these pages, the writer has avoided, so far as is possible, the use of technical terms and the discussion of controversial points, and has kept in view, not the student preparing for an examination, but one of the above-named class, and will be amply satisfied if only he can help such a person to understand better and regard with deeper interest the earth on which we live.

T. G. B.

CAMBRIDGE,
July 1912

LITERATURE

References to papers (and occasionally books) on special points in the general subject are made at the foot of the pages where these are noticed. The following list contains the titles of some treatises in the English language where the subject is dealt with more comprehensively.

AVEBURY, Lord. Scenery of England, 1902.

CAMPBELL, J. F. Frost and Fire, 1865.

CHAMBERLIN, T. C. and SALISBURY, R. O. Geology, Processes and their Results, Chapters III, IV, 1905.

DAVIS, W. M. and SMYDER, W. H. Physical Geography, Chapters IX, X, 1898.

GEIKIE, Sir A. Text Book of Geology, Book III, Part I and Book VII, 1903.

LYELL, Sir C. Principles of Geology, Chapters XV, XVII, XIX, 1872.

MARR, J. E. The Scientific Study of Scenery, 1896.

RECLUS, E. The Earth (Translation by H. Woodward), Parts II, III, 1871.

SHALER, N. S. Aspects of the Earth (section especially on Rivers and Valleys), 1890.

The Reports of the Geological Survey of America contain several important and well-illustrated essays, to which occasional reference has been made in this volume, and not a few valuable papers have appeared in the Quarterly Journal of the Geological Society of London. But of late years, in the author's opinion, geologists in North America and the younger school of them in England have often manifested a tendency to exaggerate the effects produced by glaciers and ice-sheets.

CONTENTS

LIST OF ILLUSTRATIONS

Figures 1, 2, 3, 7, and 9 are reproduced from photographs by Dr Tempest Anderson, to whom I am much indebted for the opportunity of using them. I have also to thank the Council of the Geological Society for the clichés of figures 5 and 6, from blocks used in the volume of the journal of that Society for 1871. These two and the outlines in figures 4, 12, 13, and 17 are from my own rough sketches. Figs. 11, 14, 15, 16, 18, and 19 are reproduced from various volumes of the *Cambridge County Geographies*.

T. G. B.

CHAPTER I

To destroy, to transport and to deposit is the work of rain and rivers. A stream, which has been running clear during prolonged fine weather, becomes turbid after heavy rain. If a glass vessel be filled from it and allowed to stand, the water becomes clean and a fine mud settles at the bottom. What is the source of the sediment? In part it may have been removed directly from the bed of the stream, but we soon see that there is more than this could supply, and that it must have been derived from a larger area. Again, if some gravel be heaped on a pavement we see that after a heavy shower a thin film of fine earth has trespassed on the flags. In like way, all the ground sloping towards a brook or river sooner or later sends material to discolour its waters: thus we suspect, and further examination, as we shall endeavour to show, changes suspicion into conviction, that water is one of nature's carving tools, very probably the most important. In the case just mentioned its action is mechanical, but it can also

be chemical. In a limestone quarry which has been
deserted for many years the collector of fossils
generally finds old débris to be his best hunting
ground. Here the fossils are often developed, that
is cleared from the enveloping matrix, with a delicacy
and precision, which the most skilled workman can
seldom rival. Changes of temperature, which break
the continuity of particles by expansion and contraction, and the wind which blows them away, may
have done something, but these alone are inadequate.
Another agent must have been at work, and that it is
the more potent soon becomes obvious. Go where
we will over a limestone district, we find its fossils
have been similarly developed, while this process is
much less conspicuous where the rocks are mainly
siliceous—hard sandstones and the like. Such a
development does, indeed, take place in rocks other
than limestones, but the process is a much slower
one. In ascending Snowdon from Llanberis we pass,
on the ridge of Llechog, by an old felsitic lava flow,
a relic of a volcano of which the present mountain is
a fragment. A piece of this ancient lava, broken
away from some depth below the surface, is greenish
grey in colour and all but compact in structure.
The utmost we can detect, after a careful scrutiny, is
a very faint indication of some thin streaks slightly
paler in tint than the rest of the stones. But if we
examine a surface which has been for many years

exposed to the weather, it shows a series of almost parallel whitish bands, about a tenth of an inch wide, running with slight and variable undulations, and rising for a rather less distance above depressions which are generally slightly darker in colour and perhaps a little broader. This recalls to an experienced eye the 'fluxion structure' often seen in a glassy lava, and of such it is undoubtedly a record. If, again, we examine one of those ancient volcanic breccias, which, through lapse of time, has become almost as solid as an actual lava, we find that on a specimen quarried from deep in the mass we can barely distinguish the original fragments, but that on an ancient surface they stand out more or less prominently. As homogeneous lava they would no doubt resist the continuous tapping of the raindrops better than the more heterogeneous ash which cements them, and, for the same reason, would be less exposed to chemical corrosion, which must have also played an important part in this development.

Proofs of a similar action may be found in other hill regions. To many more than Devonians, the Tors of Dartmoor are familiar—those granite crags which rise, like ruinous fortresses, above the wild rocky moorland. A nearer approach shows them to be built up of blocks which, like hassocks, are rounded at edges and corners. The same structure is also seen in the Cornish cliffs about the Land's End. One of them, in which the horizontal joints are nearer together than the vertical, presents not a little resemblance to a wall constructed with pillows. But if we examine the granite in less exposed situations, or where quarries have cut deep down into the mass, we see that the joints divide it into blocks, rectangular in shape, and as their surface becomes exposed to the sky, we can trace a gradual transition from the prismatic to the rounded outline. In the latter case we can also obtain additional evidence from the surface itself. This in a quarry is comparatively smooth, but the blocks of the Tor or cliff are rough, the grains of quartz projecting above those of felspar, which are obviously more or less decomposed. We cannot doubt that this structure also is the work of the weather; the rain, aided by changes of temperature, having slowly corroded the parts most exposed to its action.

In Cornwall, more especially, confirmatory evidence, indirect but convincing to geologists, may often be obtained. In the neighbourhood of the Land's End, we find the granite moorland strewn with fragments of a rock, so thickly speckled with black as to be distinguished at a glance from that on which it is lying. This rock is formed of two minerals, quartz and schorl[1], both very durable. But we find,

---

[1] Schorl is the name often given to a black tourmaline.

on farther investigation, that it occurs in veins, often
a few inches in thickness, traversing the granite ; from
which indeed it has been formed, by the subterranean
action of water charged with acids; a process, however, which we will not claim for either rain or rivers,
because we should find it difficult to prove that this
water had its origin in the atmosphere. But the
rain, with other atmospheric action, has detached the
schorl-rock from the granite, having eaten this away,
as already described, and left the other lying upon
its surface. The flints, which are scattered over the
chalk downs in some parts of south-eastern England,
tell a similar tale. Once they were embedded in the
soft limestone, as we can see in almost any pit, but
they are now left behind as the insoluble residue of
a mass, probably once many feet in thickness, which
has been removed by the corrosive action of water
even more completely than in the case of the granite.

But another common instance, though less direct,
is perhaps even more conclusive—that of springs in
a limestone district. In some gardens near Wells
Cathedral are ponds fed by springs, which spout up
so strongly as to form low domes on the surface of
the water. This water, as everyone will tell you,
contains much carbonate of lime in solution. We
cannot suppose it to be supplied from some subterranean factory, so it must have fallen somewhere
from the sky and have made its way underground ;

it must also have begun its journey at a higher level,
since it evidently is forced up rather than leaks out.
But 'Wells,' to quote E. A. Freeman's phrase,
'crouches at the foot of Mendips' and the Mendip
Hills are limestone.   The well-known Cheddar caves,
with their pendent stalactites, are further proofs of
this action, and a most remarkable one was obtained
about two miles from Wells some fifty years ago.
Close to Wookey Hole, noted for its remains of
hyaenas and other animals long extinct in Great
Britain, a limpid stream issues from a cave and flows
through the meadows.   Without any obvious reason
the cattle grazing in them began to sicken and the
paper manufactured at the mill on its bank showed
signs of deterioration.   An examination of the water
was made and it was found to be polluted by salts of
lead.   Further investigation showed that a lead mine
near Priddy on the Mendips, once worked by the
Romans but long abandoned, had been reopened.
Near it was a deep natural pit, into which, as a con-
venient receptacle, the refuse from the new workings
had been thrown.   That caused the mischief; for the
pit also swallowed up the rain-water of the neighbourhood and this was polluted by the lead ore in
percolating through the débris.   This fact was so
clearly established that an injunction was obtained
prohibiting the practice.

Thus water can act on mineral substances both

mechanically and chemically; but instances of the
former are much commoner than of the latter, while
we shall find that it usually operates in both of these
ways, sometimes the one, sometimes the other pre-
dominating.

As an example of the mechanical action we cannot
have a better than the so-called earth-pillars (fig. 1).
Striking specimens may be found in the Alps, though,
as we shall presently see, they are not restricted to
this region.   Some of the finest groups occur on the
Rittnerhorn, a mountain in the Italian Tyrol, one in
a valley called the Katzenbach, about an hour and a
half's walk from Botzen, and the other in the Finster-
bach, about 500 feet higher and a couple of hours more
distant.   These valleys have been excavated in a
porphyritic felstone and have subsequently been
partly filled up with morainic material—a tough clay
containing many fragments of rock, large and small.
A mountain stream has cut a glen through this clay
into the rock below and "on either side it is fringed
by the earth-pillars.   The upper part of the glen, on
the first glance, seems to be filled with these rude
obelisks, crowded like tombs in an overfull churchyard,
but, on a closer inspection, a method is seen both in
the order and in the shaping of the pillars.   Now
and then one stands alone...but the majority are
connected, and many of them form ridges of clay,
crested with pinnacles.   Each is usually capped by

a block of rock, like a turban : some, however, are bareheaded. On this block the existence of the earth-

Fig. 1. Earth-pillars, near path from Viesch to Eggischhorn.

pillar depends ; those which have lost their caps lose, not their heads only, but also their bodies. Here

and there the clay slope is furrowed by a rill, but for
the most part the 'nullahs' between the ridges and
the gaps between the pillars are perfectly dry in fine
weather." A little study makes it clear "that rain
has cut the gulleys and even furrowed the sides of
the pillars ; that the larger stones are essential to
their formation ; and that the clay becomes very
hard on drying." At first it filled the glen ; rills fed
by rain worked at its surface, and as it quickly
becomes soft when wet, they would soon plough a
number of furrows into the mass ; "one of these rills, in
deepening its bed, would encounter a boulder, and be
temporarily divided into two streams...Each of these,
after the manner of currents, would wear away the
bank of clay opposite to the stone, and would continue
to work outwards even when its bed had been cut
down below the level of the obstacle." The rills would
be separated by ribs of clay, "but as these would be
attacked by the rain, not only on both sides, but also
from above, they would gradually disappear, and the
capstones would remain exalted on pinnacles of stony
clay." The rain, also, as the pillars increased in height,
would do something, as it trickled down their sides,
to reduce their thickness, but the capstone, like an
umbrella, for a long time protects the pillar from
serious harm. The latter, however, though slowly,
becomes thinner ; "the capstone less and less firmly
supported, till at last it slips or is blown off. Then

the days of the pillar are numbered : from a pinnacle
it is reduced to a hump, and at last is wholly washed
away[1]." Similar pillars, though not generally so
remarkable as these, may be seen in several other
parts of the Alps, as, for instance, opposite to Stalden,
near Useigne in the Eringerthal, near the path leading
from Viesch to the Eggischhorn (fig. 1), and on the
north side of the Brenner Pass. The late E. Whymper
describes a remarkable group about three miles to
the south of Briançon, in which, though large stones
are frequent, capstones seem to be very rare[2]. In the
Alps these pillars are not often more than thirty feet
high, though a few of those in Dauphiné rise to sixty
or seventy feet, but on the flanks of Mount Shasta in
Northern California they may be found, according to
the late Clarence King[3], of all heights up to 700 feet.
In our own mountainous regions they are not common,
but Sir A. Geikie[4] mentions some curious examples
near Fochabers (Elginshire). These, however, are in
a conglomerate of Old Red Sandstone age, and some
remarkable instances cut from a sandstone of variable
hardness have been described by American geologists.
Here occasional lenticules, rather more durable than

---

[1] Quoted from *The Story of our Planet*, pp. 111–113, the account
being written from notes of visits in 1872 and 1880.

[2] *Scrambles among the Alps* (1871), p. 431.

[3] *Mountaineering in the Sierra Nevada*, ch. xii.

[4] *Text-book of Geology* (1903), p. 462.

the rest, produce the effect of the larger boulders in
the morainic material, crowning the pillar with a cap
which sometimes is even flatter than a turban. In
miniature, however, pillars may be found in many
places both at home and abroad, wherever a bank of
stiff clay mixed with flattish chips of stone is exposed
to the action of rain. The first which caught my
eye[1] were in the Val de Lys, above Luchon in the
Pyrenees. Most of these did not reach two inches in
height, though one giant was at least three inches.

In limestone districts many instances can be
found, as we shall presently see, where the chemical
action of rain-water far exceeds the mechanical, but
for one in which the former alone operates we must
look beneath the surface. When gravel overlies
chalk, in eastern and southern England, a pit or
cutting often displays the former filling a kind of
shaft or pouch in the latter. These are called sand-
pipes, and above one of them, the gravel which elsewhere is distinctly stratified has completely lost that
structure, the position of the fragments sometimes
even suggesting a downward movement. These pipes
also occur where the chalk comes up to the surface of
the ground, but then a little study of the neighbourhood shows that it was once overspread by gravel.
No one now doubts that these sand-pipes, which once

---

[1] In 1876. Since then I have seen them in our own country and
elsewhere.

greatly puzzled geologists, only admit of the following explanation. The gravel was deposited on an irregular surface of the chalk in which accidental depressions or cracks tended to concentrate the rain-water as it percolated downwards. As it contains some carbonic acid, it corrodes the chalk and deepens the depression, into which the gravel settles down. The process continues, and as will be easily understood, the pipe is prolonged. If the chalk contains layers of flint, these may for a time arrest the descent of the gravel, but the water makes its way through cracks and continues its work beneath them, till they break and slip down with the gravel. Thus the pipe may also contain flints unworn and obviously derived from higher positions. The chalk between Hitchin and Hatfield or between Croydon and Purley, as the railway cuttings show, is repeatedly pierced by these pipes. A yet stronger but less common proof of the solvent action of water can sometimes be found in a gravel containing many granules of chalk or fragments of shells. I remember to have seen a model, as it might be called, of a sand-pipe in a gravel pit at Barnwell, near Cambridge, and several of them in a large excavation in the shelly Red Crag near the south shore of the Deben estuary, a few miles from Woodbridge. The false-bedded Crag contained an abundance of shells, many of them in fragments. Some slight differences in the permeability of the layers had produced a local concentration of the percolating water, which had dissolved the calcareous material, obliterated the bedded structure, and thus, as it were, given a sketch of a sand-pipe. Each was some three or four feet in height and perhaps from six to ten inches in breadth.

But more usually the mechanical and chemical effects are combined, as can be seen by tracing the action of water, especially when enough to form a stream, both on and below the surface of the ground. Rivers are the makers of valleys, rather than valleys of rivers. That was not generally admitted three-quarters of a century ago, but hardly anyone would now be found to dispute it. But we reserve the history of the change of opinion, which is both interesting and instructive, for a later chapter, and proceed to trace the processes at work in the making of a valley. In the upper part of Hampstead Heath an outlier of Bagshot sand rests on London clay. In this I saw a stage of valley-making on a small scale, when walking over it the day after an unusually heavy rainstorm. Some slight irregularities of the surface had given rise to a runlet of water which had rushed down a steep part of the north-western slope, where it had cut a steep-sided channel in the sand which soon became quite half a yard in depth and a foot or so wide. Here and there was a step, perhaps three or four inches high, in the bed of the channel, and

this was its history. The sand is not quite uniform
in composition, for thin intermittent layers of it are
harder than the rest, being cemented by iron oxide
(limonite). The water, on encountering one of these,
had its down-cutting action arrested, and acquired a
plunging one at the outer edge. Thus a miniature
waterfall was formed, below which the streamlet continued its course as before. Lower down, where the
slope diminished, this tiny glen gradually broadened
out and the sand formed a small alluvial fan. We
had in fact here, as we shall presently see, a model of
what occurs on a far grander scale in many parts of
the world.

We may mention one or two other instances of
Nature's models of special forms connected with
valleys, because the circumstances under which they
occur preclude any doubt as to their origin. At the
back of Burntisland, a watering place on the north
shore of the Firth of Forth, a fragment of an old
volcano called the East Binn rises in a steep cliff.
While scrambling along its base I came to a bowl-
shaped hollow, a few yards in diameter, scooped out
in the indurated volcanic ash, its sides furrowed by
rills (then dry) which converged at the bottom and
passed out through a V-shaped notch. In its leading
features it was identical with the corries presently to
be described in mountain regions. But I have also
seen a model on a yet smaller scale. It was on

the same bank as those tiny earth-pillars already
mentioned. The leading features, small as they were—
for it was hardly half a yard across—were identical
with those in the example at Burntisland. Now in
these instances we cannot invoke any other agency
than running water, and we see what characterizes its
handwriting on the wall.

But these are not all. Early in May, 1902, the
Soufrière volcano in the island of St Vincent, after
ninety years of repose, broke into eruption[1] and
ejected an enormous quantity of scoria and dust.
To the south-west of the crater is a valley—that
of the Wallibu River—descending from the mountain
crest to the sea. This valley was filled in some parts
to a depth of 60 to 80 feet with volcanic mud and
other débris. But the old drainage lines were not
altered, and the stream, swollen by the torrential
rains, so frequent in this region, rushed down and
soon cut deep, almost to its former level, into the
incoherent débris. By the month of June the river
course was bordered by cliffs, also furrowed by the
runlets of rain, which had carved them into countless
ridges, the flanks of which were yet more closely
scarred by the rain which had fallen on them. This
can be seen at a glance in some of the excellent

---

[1] It began, after some premonitory symptoms, on May 6, but a
phase of violent paroxysms suddenly developed on the following day.

photographs[1] taken by Dr Tempest Anderson when he, with Dr J. S. Flett, visited the island at the instance of the Royal Society (fig. 2). In the spring of 1907 he returned and found that, notwithstanding the

Fig. 2. Rain furrows, Wallibu Valley.

rich tropical vegetation which had spread its protective shield over parts of the new deposit, the former valleys had been deepened and still more widened;

[1] See *Phil. Trans.* A, vol. cc. Plates 28 Fig. 1, 29 Figs. 1 and 2, 33 Fig. 2, etc.

their steep flanks of débris being furrowed by rain-rakes, even more conspicuously than on the former occasion[1]. But the same lesson is taught by the cliffs of glacial drift on some parts of our eastern coast, by the Tertiary marls on the western flanks of the Apennines and by similar material in parts of the United States, such as the Bad Lands of Nebraska ; numbers of little ravines running with almost parallel courses down the nearly precipitous cliffs, but now and again effecting a junction and cutting deeper into the slope. Can we then apply this explanation to the corries and cirques, the glens and gorges, which occur, sometimes on so grand a scale, in many mountainous regions? We can more conveniently discuss that question after considering some instances where the chemical often exceeds the mechanical action of water.

In many limestone districts (for in them the solvent action of the rain is much more conspicuous than elsewhere) the surface of the rock is often curiously pitted, grooved, scarred and barren. These features are common in the Carboniferous limestone of the Dale-land in Western Yorkshire, in many parts of the Alps, especially in the Karst region of Carniola, and in the noted Steinerne Meer in the Salzburg Alps. The last is a plateau about five miles in length and fully two miles in breadth, which is rather more than 6000 feet above sea-level. A more dreary

---

[1] See *Phil. Trans.* A, vol. ccviii. Plates 9–12.

spot could not easily be imagined. Here and there, but very seldom, is a ragged *latschen*[1] or other blasted pine, ill-grown, twisted by the wind, perhaps dead, but the region, except for a rare and faint streak of green or a few stunted and starveling alpine plants, is a wilderness of rock, weathered into myriads of holes of the strangest form, ramifying into shallow gulleys and often ending in pipes or narrow shafts which pierce vertically into the mass. Neither rills, nor streams, are visible ; the melting snow and the summer rain are quickly swallowed by these innumerable mouths. Similar instances of water channeling can be found in the neighbourhood of the Gemmi Pass and other limestone districts of the Swiss or Savoy Alps, but nowhere are they better or more easy of access than in the lake region to the south of Ischl. But some parts of England, especially in Western Yorkshire, afford a striking object-lesson of the solvent power of rain. Here and there, in a hilly limestone district, large blocks of hard grits have been dropped, during the glacial epoch. They must have been deposited on its surface : they now stand on pedestals from about twelve to twenty-four inches above the present one ; these indicate the thickness of the layer which has been removed, and that by subaerial action alone, during the interval, whatever

---

[1]   A kind of dwarf-pine (*P. pumilio*).

may have been its length, between the disappearance of the ice and the present day (fig. 3).

One more instance may be given of the power of running water, in this case largely supplied by melting

Fig. 3.   Erratic Block, on pedestal.   Norber.

snow. Professor Garwood has described[1] how, in Spitzbergen, a level plateau, of rather incoherent materials, was capped with snow and terminated in a rather

[1] *Geographical Journal*, vol. xxxvi. (1910), p. 310.   Pl. I fig. 1.

steep slope above a lowland by the sea. The snow
had a protective effect on the plateau, but in the
summer season it fed numbers of rivulets, differing
little in size, which made their way down the slope,
and caused it to look as if it had been furrowed by
a gigantic rake, for they generally followed an independent course from top to bottom.

We may now trace the underground course of the
water swallowed up on these surfaces of limestone.
In sundry districts, including those already mentioned, a stream contrives to form and run for some
distance before it is engulfed. The 'pots' or 'swallow holes' of Western Yorkshire have long attracted
notice, and there is no more striking or instructive
instance than Gaping Ghyll, which is at a height of
nearly 1350 feet above sea-level on the western flank
of Ingleborough.   Here, where the Carboniferous
Limestone is exposed at the surface, it is carved into
basins, channels and pipelets, but it is sometimes
covered by a few feet of drift.   In one of these areas
a little stream—the Fell Beck—has cut down to the
limestone.   On either side the banks slope for some
yards down to the water.   But looking along its
course we see that they curve round so that the
valley ends, though in the wrong direction.   A few
steps more and we have the explanation ; the water
plunges into an open shaft about six yards wide
and five yards across.   This shaft was descended in

August, 1895 by Mr E. A. Martel[1] with the aid of a rope ladder at the end of a stout cord. At the top is a small side opening which leads to an irregular shaft ultimately joining the main one. Professor Hughes had already sounded this, and found the depth to be 330 feet, a measurement confirmed by Mr Martel. The falling water afforded him an abundant and increasingly vigorous shower-bath, which, however, did not cause more than inconvenience. On approaching the bottom the explorer found himself suspended in a large open space, and on reaching the ground found that the shaft ended in a 'vast hall, 500 feet long, 80 to 100 feet high, 66 to 116 broad.' It was blocked at either end by fallen rock, the water escaping through the eastern mass of débris[2].

Some years before Mr Martel succeeded in exploring Gaping Ghyll it had been discovered what became of the engulfed water. In the rocky flank of a valley, near the village of Clapham, rather more than a mile away in a straight line, are the noted Ingleborough caves. These evidently have been cut by a stream, which emerges a short distance farther up the valley, after working its way through an exceptionally hard bed of limestone, the presence of

[1] E. A. Martel, *Alpine Journal*, xviii. p. 120, and *Irlande et les Cavernes Anglaises* (1879), ch. xxiv.

[2] According to Mr Martel the shaft has been driven through the Carboniferous Limestone down to the underlying Silurian.

which no doubt led to the formation of the caves. The connexion with Gaping Ghyll was proved, not only by throwing chaff and colouring matter into the water which plunges into it, but also by the help of nature. Several years ago a 'cloud burst' on Ingleborough changed the stream into a torrent; in due course the water at the outlet in the valley increased in volume till it flooded the upper cave, resuming for a time its ancient channel[1]. In not a few caves water springs from openings in their walls can be followed for a time, and then passes out of sight on an independent journey, or a dry cave joins another one down which a river is flowing.

Caves and swallow holes abound in the Carboniferous Limestone districts of England, Ireland, Belgium and the United States, some of them, especially in the last country, far exceeding any known in Great Britain. The limestones also of Secondary age, in many parts of the Alps, and still more in the Causses of France (the plateau region drained by the Tarn, Lot and Dordogne) afford numerous caves and subterranean watercourses. In every mountainous region where water is swallowed up on the higher ground it may break out on the face of a cliff or at its base in the bed of a valley below. The

[1] See W. Boyd Dawkins, *Cave Hunting*, ch. ii. for other instances in England. The distance between the farthest point reached in the caves and Gaping Ghyll is about 4000 feet, the fall being about 130 feet.

annexed diagram (fig. 4) shows how local conditions have modified the form of the culvert, which has now become dry because the stream has sought another path ; but instances of streams issuing from the rocks in fairly full vigour are too many to enumerate. The Sorgues at Vaucluse near Avignon, the cascades

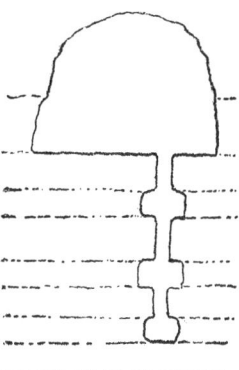

Fig. 4. In face of cliff, near Untersberg, Salzkammergut.

about the Drei Brunnen at Trafoi, in the Tyrol, the Sieben Brunnen of Lenk in the Western Oberland, the Orbe and the Loue in the Jura[1], are all instances of the emergence of water, which has been elsewhere swallowed up, and perhaps joined by other streams

[1] E. Reclus, *The Earth*, ch. xliii. E. A. Martel, *Les Abimes*, p. 437. G. F. Browne, *Ice Caves*, pp. 59, 122.

during a subterranean journey. The region of the
Karst, which has already been mentioned, is pierced
by funnels and riddled with caves. The most remarkable instances of the latter are those near
Adelsberg[1]. To the north of the town the river Poik
enters a portal in a limestone crag and can be followed
for more than a thousand yards, being joined near
the entrance by the famous caves which are now
dry beds of that river or of tributary streams. The
Poik, after a subterranean course of about twelve
miles, during which it can be approached in one
place down an opening from above, emerges under
another name, the Planina, but their identity is
beyond reasonable doubt. About a dozen miles to
the south-east of Adelsberg limestone cliffs surround
an extensive and generally level basin. The floor
of this is pierced by funnel-shaped holes about 400
in number. These, for the most part, are dry, but
water rises up into them after heavy rain, and sometimes actually overflows in such abundance as to
change the plain for a time into a lake, the area of
which has been known occasionally to be nearly
25,000 acres. There must be a number of subterranean channels in the limestone rock, communicating
one with another and forming a regular drainage
system, which becomes so gorged with water as to be
unable to carry away all of it ; thus the rest ' backs

[1] E. Reclus, *The Earth*, pp. 295-299.

up' the shafts and overflows the surface which they generally keep in a very dry state.

In all these swallow holes and caves the erosive action of the water is largely, and in some cases mainly, chemical, but we must not forget that wherever it falls from a height or is running rapidly, it exerts a mechanical force, and this, in such cases as Gaping Ghyll or similar shafts, must increase with the depth till at last it may exceed the other one. The fact, however, that natural shafts, swallow holes, and caves are very rare, if not wanting, in all but limestone regions, proves that chemical action is the more essential condition for their occurrence.

One effect, on the whole mechanical, of water when percolating beneath the surface must not be forgotten, because it is sometimes locally important. After continuous wet weather the ground becomes saturated to a considerable depth, the coherence of the materials is diminished and on sloping surfaces landslips may occur. This often happens on the banks of railway cuttings in the London clay, though they are protected by herbage and sometimes strengthened with burnt material. When the mass has been 'lubricated' to a depth of two or three feet, the turf above begins to bulge ; then it cracks, and finally slips down. After heavy rains these slides have sometimes temporarily blocked the lines. The same thing happens on a larger scale when a permeable

stratum ending in a cliff rests upon an impervious
one which slopes outwards, as in the case of the
famous landslip which occurred, after a very wet
season, on Dec. 24, 1839 between Lyme Regis and
Axmouth. Here a mass of chalk ending in a cliff
above the sea passes down into a less coherent
calcareous greensand, beneath which is a shaly clay
of Liassic age—all sloping seawards. The last-named
bed arrested the downward passage of the water,
which saturated and loosened the greensand until
a broad strip of the overlying chalk, more than
a hundred feet in thickness and three-quarters of
a mile in length, slid down seawards, breaking up
into rudely prismatic blocks.

In mountain regions, where the rocks are often
tilted at high angles and the valleys cut deep into
the successive strata, these landslips are on a greater
scale and far more terrible. Such, for instance, was
the historic fall of the Rossberg in Switzerland. This
is a mountain rising some 3500 feet above the valley
between it and the well-known Rigi. Permeable rocks
overlie, with an outward slope, comparatively impervious strata, and here, after an exceptionally rainy
season, a great mass of conglomerate, some three miles
long, 350 yards broad and about 100 feet thick, slid
down to the valley on Sept. 2, 1806, breaking up into
fragments as it went, and destroying about 94 acres
of fertile land and three villages with part of a fourth,

thus bringing death and burial to 484 persons. Not a few similar cases, where water has added catastrophic to its ordinary results, have occurred in the Alps and other mountain chains.

CHAPTER II

THE MAKING OF VALLEYS

WE have endeavoured to prove in the preceding chapter that water in its liquid form is a powerful agent in earth sculpture. It is aided no doubt by heat and cold and the wind, it may act in the form of ice or of the sea, but each of these produces its own results, and the denuding effects of rain and rivers can be appreciated by tracing the history of valleys. These also can be found in miniature and the processes of making be watched.

When the water is draining off from a gently shelving sandy-shore, just exposed by the receding tide, we find the level of the surface quickly broken by a number of channels. They begin sometimes almost imper-ceptibly, but quickly develop into tiny watercourses with rather steep walls and flat beds. The walls are steep, because the material is homogeneous and fairly coherent ; they are rather low and the bed is flat

rather than angular in section, because the coherence
soon reaches its limit. At first these streamlets are
fairly numerous, but they tend to unite on their
downward course, because if any one of them be
diverted by some chance obstacle, it will be directed
from a straight course towards one of its neighbours
and in some cases it may take an oblique path from
the beginning. Thus a map of these watercourses
often resembles a drawing of a leafless twig of a beech
tree or of willow roots when spread out in water.
But as the slope steepens, the number of these
tributaries diminishes, their length increases and
when we pass from this gently shelving floor to a
bank of steep stony débris, we find that the many
twigs pass into one strong stem with few and insignificant off-shoots. Thus a river system, to give a
very rough comparison, may present a variety of types
from the growth of a beech to that of a Lombardy
poplar, the latter one being common in any mountain
region. We mentioned their occurrence on a small
scale in close furrowed slopes of mud and débris, and
they are exhibited on a larger one by almost any
steep slope in such a region as the Alps. Sometimes,
as will presently be described, their beginning is a tiny
corrie, but more often a gradually enlarging furrow,
at first barely perceptible, makes its appearance on
the slope, without being joined by any tributary
large enough to be noticed from a distance. The

section of the furrow, so long as it is cut through débris, is V-like, because its flanks are not sufficiently coherent to remain in an upright position, but if it reach the underlying solid rock, the form of the channel will be changed, the walls become steeper and the groove more of a gash. To these and other modifications in form we shall return, for the moment it is enough to notice that in all cases the brooks gradually unite to form a river, and the rivers may combine in a great trunk stream.

Suppose then land is beginning to rise from beneath the sea, as a low dome. Assume this, for the moment, to be circular in plan. No sooner is it exposed to the sky, than the rain will fall upon it and will begin the downward transport of its materials. These possibly may be removed so uniformly that the rounded outline will be retained and the dome be gradually lowered till at last it is hardly more prominent than a shoal when bared by a falling tide. But such uniformity would be exceptional. Usually the sculpturing process will begin at once, and little river systems be initiated, as above described. These will radiate outwards, to the different points of the compass, and as elevation continues, will obviously be less and less likely to unite. Such an area would exhibit a number of rivers about equal in magnitude. But a much more probable form for the rising land is a very elongated ellipse, to the minor axis of

which the majority of the larger trunk streams
would be parallel. Another factor will almost
certainly enhance this inequality of development.
Before the crest of the dome emerges from the sea
it will be attacked by the waves, which will plane
away its surface, and if it is formed of stratified rock,
the area finally uplifted will consist of a series of
elliptical shells, the materials of which differ in their
powers of resistance. Suppose them to be alternately
hard and soft. In the former the V-like section will
be preserved, but the latter yields more easily and
another modification appears. The rain, which has
fallen on this portion, will soon be checked in its
downward course by the outer band of hard rock,
and will wander over the softer material, making
some sort of a channel in this until at last it meets
with one of the radial grooves. This develops a
second system of valleys, making a high angle with
the others, and bounded on either side by the outcropping harder rock.

Thus the valley courses in an elevation of this
nature form two groups, the one following the dip, the
other the strike of the beds, and named from that
dip valleys and strike valleys. Even crystalline rocks
not unusually show a system of divisional planes,
more or less resembling stratification, with a jointing
perpendicular to them, so that the terms dip valleys
and strike valleys may also be extended to them, but

we may avoid any difficulty of this kind by calling
them, where speaking of them inclusively, transverse
and longitudinal valleys[1]. Complications will be
introduced by inequality in the uplift, by its occurring,
with apparent independence, in different places, and
by actual movements in the opposite direction, but
to these we may return ; for the present it suffices to
say that this method of river formation is the key
to the physical structure of any area of the earth's
surface, be it large or small.

The beginning of a valley assumes more than one
form. Sometimes it starts from a rather bowl-shaped
head; sometimes a very faint notch gradually becomes
perceptible on a slope, broadening and deepening on
its downward course. The former is due to the con-
currence of a number of streamlets, of which none,
when the hollow is a small one, is strong enough
to mark its path by a furrow and which unite, owing
to some favourable circumstance in the original
configuration of the ground, almost at the same
point, and then have sufficient power to cut a notch.
The exact outline—the slope of their sides—in both
the one and the other, depends on several conditions,

[1] Some authors of late have preferred to call them 'consequent'
and 'subsequent' valleys, but the multiplication of terminology
seems to me an evil which in this case has not the compensation of
greater simplicity and explicitness. 'Transverse' and 'longitudinal'
were the terms used by J. B. Jukes to whom we are indebted for a
most important paper on valley-making. See page 132.

but notably on the nature of the rocks which are being carved. If these are fairly hard and homogeneous, the walls will be steep ; if soft, they will be gentle, if variable they may be in steps. Among our English chalk downs the heads of the valleys very often are more or less bowl-shaped, and here the slopes of their sides become rather steeper, but sometimes they seem to die away almost imperceptibly among the hills without any attempt at a notch. This is because chalk is soft, uniform in texture (flints of course excepted) and readily yields to the corrosive action of the atmosphere. In our more mountainous districts both types of valleys occur, but perhaps that starting with a faint notch is the commoner, and the outlines throughout are of a bolder type.

These bowl-like hollows often become larger as the hills increase in size and their walls steepen in the more durable rocks. They are then called corries and cirques—the latter term being reserved for those which have the steeper walls—and are due to the action of rather numerous convergent streams in rocks which, though strong enough to maintain fairly vertical faces, yield rather readily to denuding agencies. Some writers have attempted to draw a distinction between small cirques, which they call alcoves, and large cirques, and to associate the names with a difference in origin. This distinction appears to me, after an experience of more than forty years, to have no valid foundation. As with the corries, so with the cirques, you may pass from small to large, from the pigmy to the giant. I have seen a model cirque as well as a model corrie. It was in a large pit in the Bunter pebble-bed, worked near the Cannock and Rugeley railway (Staffordshire). Here, after an unusually heavy rainstorm, a semicircular recess had been cut in one place downwards from the surface. Its walls were vertical ; perhaps about four feet high (the diameter being rather less), and its bed like half a saucer. Here the streamlets had not made any perceptible furrows on the side, probably because the pebbles compelled each one to follow a rather zigzag course ; but in the larger cirques the furrows are often plain enough.

The following quotation[1] describes two which may be seen from the rugged pastures of the Blacken Alp (5833 feet) on the way from Engelberg to the Surenen Pass. They are cut out of the highest part of the Uri Rothstock *massif*, and separated by a spur named Rothschutz on the Federal Map and are very similar in appearance, though the eastern one, on the whole, is the finer of the two (fig. 5). "Its other extremity is the Blackenstock (9587 feet) and from this summit to the Rothschutz (9278 feet) runs a line of crags not much inferior in height ; the chord joining their two ends is about 2800 yards long, and the sagitta of the arc,

[1] The author, *Quart. Journ. Geol. Soc.* vol. XXVII. (1871), p. 314.

about 650 yards : but the spurs projecting from the
two extremities give a more semicircular appearance
to the cirque than these measurements would suggest.
Above the usual taluses of débris rises a high band
of cliffs of a hard yellowish limestone, which supports

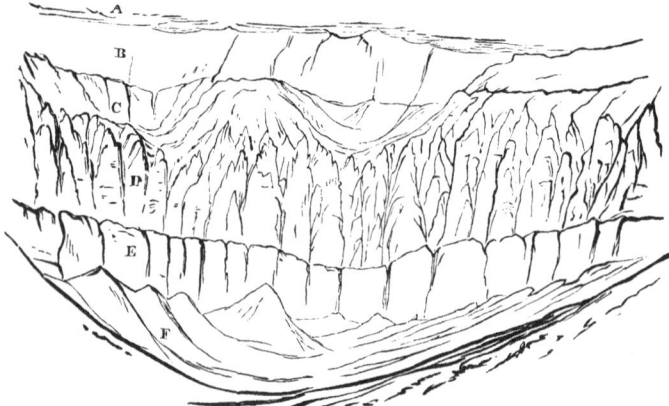

Fig. 5. Cirque in Rothstock. A. Clouds concealing peaks. B.
Limestone cliffs. C. Shaly slope with small corries and
snowbeds. D. Shaly cliffs furrowed by streamlets. E. Limestone
cliffs slightly grooved by these. F. Floor of cirque with talus
heaps on side.

a still loftier belt of a reddish rock, doubtless a rather
sandy and coarse calcareous shale : over this is a sort
of terrace-shelf or slope, hollowed out into small
combes [corries]; and then rises another barrier of
limestone cliffs forming the lip of the cup-shaped

hollow. Clouds prevented me from seeing the sky line in more than one place, but it is nowhere more than a few hundred feet below the peaks named above. For the same reason, I cannot attempt to give any accurate estimate of the height of the cliffs; but the two lower stages appeared to me together not less than 500 feet. The strata lay tolerably horizontal, only curving upwards somewhat in the western part of the western cirque. The most remarkable thing about the cliffs was the belt of reddish shaly rock, which was furrowed by a vast number of little gorges —which were often only a few yards apart and occasionally united—so that this part of the cliff really looked like a badly ploughed field set up on end. Down these gorges, many of them dry in August, little rills descend from the snow on the ledges and in the combes above, which have generally made some trace, corresponding with their size, on the harder limestone below—sometimes a mere stain, sometimes a well-marked groove."

Still larger cirques may be seen in other parts of the limestone Alps, such as Am Ende der Welt at the head of a glen to the north of Engelberg, the Fer-à-cheval near Sixt and the magnificent Creux de Champ in the Diablerets near Ormont Dessus. Somewhat smaller are the Croda Malcora cirque[1] in the Sorapis

---

[1] *Id.* p. 313. I went up into it in 1872 and was satisfied that though inferior to Gavarnie it was a true cirque.

above Cortina d'Ampezzo in the Dolomites, and that
of the Creux du Van in the Jura. The Pyrenees also
afford several examples, the best known of which is
the famous Cirque de Gavarnie. But among limestone mountains, they may be found of all sizes from
such as these downwards. The annexed engraving
(fig. 6) gives a rough sketch of a small cirque, high
up in the limestone cliffs, on the left bank of the Aa
valley, a short distance from Engelberg. It is an
exceptionally good illustration of the genesis of a
cirque, for here six or seven small streams issue from
the rock and have worked out a hollow, only a few
dozen yards wide, yet of the true cirque type. It
would be easy to quote other cases, where small
cirques or corries are developed on the face of
great walls of limestone rock, in a position where it is
impossible, so far as I can see, to invoke any other
agency than that of small streamlets, fed by springs
or by little gatherings of snow on ledges above.

Among crystalline non-calcareous rocks, cirques
are far less common. Here they generally take the
shape of corries; that is, their walls are not so often
vertical, and their outlines a little less regular, since
the materials generally are not so favourable to the
production of such numerous streams or of vertical
precipices. Still cases may be found, as for instance,
high up on the eastern side of the Grand Combin,
where the névé basin of a glacier is enclosed by a

sweep of cliffs, which, did the ice melt away, might well be claimed for a cirque. The situation of a cirque or a corrie, it may be well to remark, is not always at the head of a valley. The Fer-à-cheval,

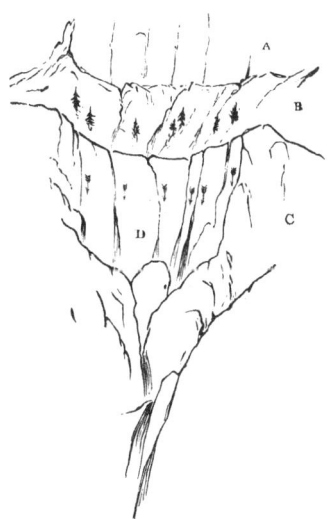

Fig. 6. Small cirque near Engelberg. A. Limestone cliffs. B. Shaly bank with some trees etc. out of which the streams break. C. Limestone cliff. D. Cirque. ↓ Cascades.

for instance, and the two cirques under the Uri Rothstock are both at the side of a well-marked valley so that it is extremely difficult to understand how glacier ice could exercise that 'sapping' and

' plucking ' effect which has often been invoked of late years, but which I believe to be largely imaginary.

Cirques, in short, require for their development, unless I wrongly interpret the evidence, the co-existence of the following conditions : (1) several small streams, which are fairly constant and have a tendency, from the shape of the ground, to be confluent, (2) strata, moderately horizontal, over which these streams fall, and which by their constitution yield considerably to the other forms of meteoric denudation, (3) these strata must nevertheless allow of the formation of cliffs, and thus perhaps the most favourable structure is thick beds of limestone with occasional alternating bands of softer rock.

Gorges are perhaps the most impressive forms produced by strong and rapid streams acting on durable and homogeneous rocks. Thus they are frequent in mountain regions ; becoming smaller among hills, and vanishing as these subside into plains. In the Alps they are common, especially where the velocity of a stream is suddenly increased by descending a step in the bed of a valley. Here, for instance, the floor of a lateral glen at its junction with a main valley is often much higher, perhaps some hundreds of feet, than the latter, so that a steep almost precipitous descent leads from the one to the other. Opinions differ as to the exact explanation of these hanging valleys, as they are called, but, whether the larger river has deepened its channel more rapidly than the smaller one, or glaciers in the two valleys have produced the same effect, a gorge, like a great notch, is almost always found in the lip of the above named rocky step. They occur alike in the limestone and the crystalline districts. Among the former are the gorges of the Rosenlaui and the two Grindelwald Glaciers. That of the Unter Glacier, a magnificent example, was hid beneath the ice half a century ago, but it was exposed by the rapid shrinkage of the glaciers which began soon after 1860, and it has since then been made accessible. Its head is masked by ice which still covers the upper part of the last step in the bed of the mountain valley, but fine views of the rest can be obtained from above and from a gallery constructed by the side of the torrent. The gorge is a few yards wide, its limestone walls rising almost vertically, and retaining in many places the indications of torrent action far above its present level, in some cases to the top: such as scooped-out hollows of all sorts and even pot-holes, often partly cut away as the channel was lowered, till at last the gorge is choked by ice from the glacier above. But similar gorges, which no doubt were begun, if not sometimes mainly formed, in the same way, occur in many parts of the Alps. Such are those at the entrance of the Eringerthal and of the Einfischthal, that of the Trient at Vernayaz (fig. 7) and of the Tamina at Pfäfers:

those on each arm of the Visp near Stalden, the one
cut by the torrent from the hanging valley of Saas
Fee, and a host of others, small and large, which it is
needless to enumerate. These, however, are made by

Fig. 7.   Gorge of the Trient, Vernayaz

torrents largely fed by glaciers, the grit and mud
from which must greatly augment the cutting power
of the water, and when they occur in such districts as
Wales, Lakeland, or the Scotch Highlands, they may
be regarded as being, to a great extent, relics of the

Ice Age. In fact, though gorges may be found in regions where even permanent snowbeds are not likely to have existed, they are rare in occurrence and small in size.

One kind of gorge demands a brief notice, though the precise explanation of it has been much disputed. On this question, however, we need not dwell, because it does not affect the main point, that the gorge was cut by a torrent. This kind, however, differs from the others in traversing a barrier which crosses a valley. That of the Kirchet is the most noted instance. Here the bed of the Aar valley, between Imhof and the level plain extending by Meiringen from the lake of Brienz, is crossed by a limestone barrier over 300 feet in height, through which the river has cut a path, seldom more than a few yards in width, and at the lower end so narrow in one or two places that a man may almost touch the walls with outspread hands. Here also, the lower parts are waterworn in the usual way, and no doubt the marks once extended up to the top. A century ago this gorge would probably have been regarded as a rift produced by some convulsion in the earth's crust, but a closer study manifests the improbability of that explanation. The longer and lower part of the gorge is now made accessible by a slight gallery of wood. Visitors, who have reached the end of this, can vary the expedition by ascending a steep ravine which

leads to the upper part of the Kirchet barrier. On
the sides of this, also, he will recognise the signs of
water action. Down it a tributary stream, fed by the
glacier which once passed over the Kirchet, but has
long melted away, must once have plunged. It has
no resemblance to a crack, but is identical with
dozens of lateral gullies on the flanks of ordinary
ravines. If then it has been the work of a stream,
we need not hesitate to ascribe the other one to the
same agent.

Gorges of this kind are not very numerous in the
Alps, though, as we shall presently see, narrower and
broader parts often alternate in their valleys, but
there is a fine example, that of Sottaguda, near
Caprile in the Dolomites. The Rhone also at St
Maurice cuts through a limestone barrier which runs
across the valley, and it is noteworthy that both here
and at the Kirchet this barrier is not severed quite
at its lowest point. On that fact, however, for
which more than one explanation has been proposed,
we need not linger, since, whatever may have been
the directing influence, we cannot doubt that a torrent
has been the agent which has severed the barrier. In
fact, as we wander up and down the valleys in the
Higher Alps or Pyrenees—we mention these because
the 'writing on the wall' is in bolder characters than
in our own country—we find that where a stream
descends a step, whether in a lateral valley or at the

junction of one of these with a main valley, it makes
its mark upon the edge. This, if the stream be
small, and especially if it be clear, may be hardly
perceptible, but as its volume and its turbidity increase, or in other words as it is more largely supplied
from glaciers, its channel becomes a groove and the
groove deepens and extends backwards till it becomes
a gorge. The torrent, which carries down to the Visp
at Saas the drainage from the glaciers in a gigantic
corrie—the peaks of which extend from the Mittaghorn
to the Nadelhorn—flows with little disturbance over
the comparatively open floor, but, immediately after
leaving Saas Fee, it plunges down a ravine in the
rocky step at the mouth of the hanging valley. That
village and Saas Grund in the main Visp valley are
separated by a rocky wall about 800 feet in height,
into which the torrent has sawn a channel. The
bed of this slopes steeply, for the torrent emerges on
a level with the main river.

The depth of one of these gorges at the end of
a hanging valley depends on the erosive power of
the torrent. Here, where it is large and its waters
turbid with débris from the glaciers, the gash is on
a grand scale, like several which have been already
mentioned, but we can demonstrate the importance
of both conditions without quitting this part of the
Vispthal. On its opposite side, within a distance of
about four miles, three torrents emerge from hanging

valleys, all terminating in a well-marked rocky step. In the southernmost, and the middle one—the valleys leading to the Antrona and the Zwischbergen passes —the glaciers are unimportant and in each the lip of the step is but slightly notched, but in the northernmost, where the glaciers are rather larger, the notch is a little more deeply cut though it is in no way comparable with the great gorge on the other side of the Vispthal.

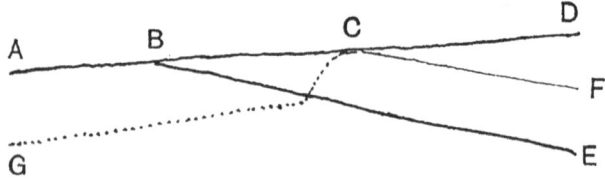

Fig. 8.   Cutting a step (the dotted line).

Waterfalls are also instances of the cutting power of a stream.   Illustrations on a small scale are frequent in a hilly or mountainous region.   As a simple one suppose a bed of hard rock (*BECF* in the diagram fig. 8) to crop out on a smooth surface *ABCD*, shelving slightly towards *A*, between two softer beds *D* and *G*, and a stream to begin moving along the line *DCBA* in the direction of the lettering.   The bed *BECF* would be worn away more slowly than those above and below it.   In the upper of them the stream could

not appreciably deepen its bed, but as soon as it
encountered the softer rock beyond B it would begin
to cut down into it. This would set up a slight
plunging motion in the water, and initiate a step
protected by the outcropping edge of the harder
rock. The result would be a slight increase in the
velocity of the stream above the step and a corresponding one in its erosive powers. The whole
thickness of the harder bed would be exposed together
with a greater amount of softer rock beneath it. The
backwash of the water would begin to work upon the
latter and undermine the ledge: pieces of that would
fall, and the cascade recede upstream (fig. 9). Good
illustrations of this can often be found on mountain
slopes at the outcrop of alternating banks of limestone and shale, and it may be worth noting that in
cases where that outcrop makes an oblique angle
with the course of the stream it not only comes down
in a series of little steps (the height and breadth of
each depending, of course, on the thickness of the
beds), but also works away to one side—that towards
which they are sloping—so that the section of the
furrow instead of being a regular V has the steeper
arm on that side. Many waterfalls have been formed
by that undermining process, and even Niagara has
been caused by a thick nearly horizontal bed of limestone (about the same age as the Wenlock limestone
in Britain) which overlies one of more perishable

materials. Here, everything is on a grand scale, but we may trace the process through any number of graduated examples down to the rivulet as it tumbles over a tiny cliff but a few inches in height.

Since 1678, when the Falls of Niagara were seen

Fig. 9. Hardraw Scar, Yoredale.

for the first time by a European, Father Hennepin, they have receded for some distance up stream and have changed considerably in form. The extent of the latter is difficult to estimate for his rude picture so greatly exaggerates their height that a comparison

of the details is difficult, but there are no signs either
of the great 'horseshoe' on the Canadian side or of
the deep notch by which of late years its outline has
been injured. For nearly a century after that date
Niagara was seldom visited. But now the Falls have
been carefully studied for fully seventy years; and
the results published by observers both in Canada
and in the United States. Dr J. W. Spencer has
given an elaborate summary of the results of work
done by himself and others in a volume entitled
The History and Geology of the Falls of Niagara[1].
When he wrote the height of the Canadian Fall,
which is slightly the lower, was 158 feet, and
the width of the gorge immediately below was
1200 feet. Between 1842 and 1905 the average
rate of recession had been 4·2 feet a year, but the
amount was decreasing, and for the last fifteen of
those years had been reduced to 2·2 feet though in
the preceding fifteen it had been 4·4 feet. This
change is probably due to a marked reduction in the
volume of water which passes over the Falls, so much
having been abstracted of late years to work
machinery, especially for conversion of the water
power into electricity. By this abstraction, as Dr
Spencer states, the beauty of the Falls has been
impaired. The depth of the water in the neighbourhood of Goat Island has already been reduced by

[1] Published by the Geological Survey of Canada, 1907.

over a foot with the result of laying bare rock in the
shallower part, and he anticipated that at the present
increasing rate of diversion the American Falls would
be reduced from a sheet of a thousand feet in width,
to a few narrow streams coursing down the deeper
channels, and the depth of water on the Canadian
side lowered by two or three feet[1]. " Niagara Falls
reached their supreme magnificence by 1900 when
the perimeter of the Canadian Falls was 2950 feet
and the American 1000 feet. The whole is now
reduced by encroachment to about 3500 feet. With
the use of the full franchises the entire width of the
falls will be reduced to 1500 feet and then they will
be wholly in Canadian territory, except small streams
coursing down the ancient river bed over the Goat
Island shelf and the present route of the American
channel and falls." Thus, with the twentieth century,
Niagara has been appropriated for the production of
wealth, but has ceased to have value as a geological
time-piece.

More than one estimate has been made of the
number of years which have elapsed since the Falls
began to carve a channel in the limestone plateau,
beginning at the edge of the escarpment at Queens-
town about seven miles below their present position.
At first, according to Dr Spencer, their height was
only about 35 feet and their channel very much

[1] *Loc. cit.* p. 267.

shallower. Now the depth of their bed, as near as it can be measured to the foot of the cascade, is from 80 to 100 feet. The history, as he relates it, is a rather complex one, and the result has been much variation in the time-estimates, as the several stages in the process have been recognised. If the recession had been uniform and at the rate of four feet a year, the beginning of the Falls would have been 9000 years ago. Bakewell in 1829 had allowed 12,000 years; Lyell in 1841, by taking a slower rate of recession, gave it as 35,000, but greatly reduced this amount after obtaining more precise estimates of the rate of change. Dr G. K. Gilbert in 1886 put it so low as 7000 years, by using for the calculation the maximum amount of recession, the mean rate of which would give 9000 years. Dr Spencer, making allowance for a much slower rate in the earlier stage of recession, argues in favour of a considerably larger figure, a total of 39,000 years, from which, if the lowering of the Erie basin—an early stage—be more rapid than he had estimated, 3500 years must be deducted. Thus Lyell's original estimate, which was little more than a conjecture, corresponds very closely with the results of an elaborate investigation. It would, however, be premature to say that the controversy is at an end, but we may venture to assume that the age of the Falls as a whole is not greater than the figures just quoted. As this work

has been carried out in post-glacial times, it gives
some help in any effort to approximate the date at
which the Ice Age came to an end.

The form of the cross-section of a valley next
calls for attention. If the walls on both sides of a
gorge are practically vertical, it must be cut in rock
which is not only hard but also uniform in power of
resistance. If, however, that consists of horizontal
beds of harder and softer material, the latter yield
more readily to meteoric agencies of denudation;
thus weakening the foundation of the former, and
causing pieces of them to fall. Then, instead of a
vertical wall we have a series of crags alternating
with slopes, a structure which is usual in the valleys
of mountain districts carved out of suitable sedimentary rocks, and may also be seen, though generally
less conspicuously, even among those which are
crystalline, for they also, as already said, not seldom
have a rather similar divisional structure. If the
outcrop of inclined beds be generally parallel with
the direction of the stream, the sides of the valley
will slope at different angles, for, whether the rock
be sedimentary or crystalline, it is usually affected
by a system of joints, more or less perpendicular to
the bedding or its equivalent, which divide it into
blocks. The equilibrium of those on the one side is
comparatively stable for, if slightly disturbed, they
are brought back by gravitation to their former

position, but on the other it is the reverse, so they are readily displaced (fig. 10).

Hitherto the channel of the stream has been assumed to be a straight line. In that case its rate of flow is greatest in the middle, and is the same on either side, but if the channel be curved the velocity is greater near the concave than the convex bank.

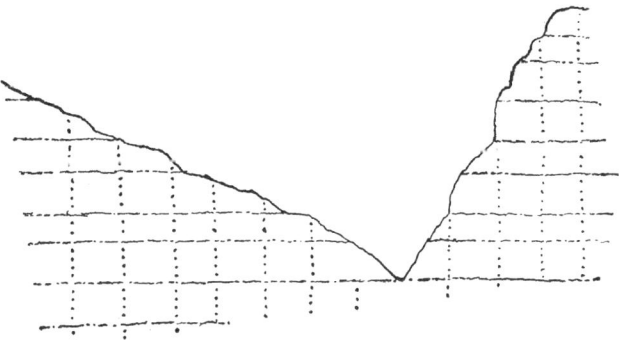

Fig. 10. Effect of bedding and joints (dotted lines) on form of valley.

The former will therefore be worn away more rapidly than the latter, on which also the bed will 'shallow' more slowly. Thus the process of erosion keeps the concave bank the steeper of the two, and the whole valley not seldom presents that form. Again as the slope of a valley-bed diminishes, which, as will be seen, it usually does, the velocity of the stream is diminished

though the volume of its water is increased by the
incoming of tributaries. The result of this is a
broadening and other changes of form in the valley ;
its banks slope less steeply ; its course becomes more
winding (fig. 11). So far as this is concerned the form
of its section is that of a trough instead of a V, in

Fig. 11. The Kent, below Kendal.

which, however, the two sides have a different slope.
As the course of the stream is now more easily diverted,
it trespasses here and there on its banks, and no longer
occupies the whole breadth of the valley. Thus the
trough section, often wider than deep, characterizes
the valley made by a large strong river. For instance,

where the Meuse passes through the older Palaeozoic rocks of the Ardennes, as between Revin and Deville, its valley winds, sometimes considerably. In places there is little or no level land by the river, the bank shelving down towards its concave edge, and descending steeply to the convex one, but in other places there is a moderate extent of flat meadow in which the channel takes an independent course. Yet here also the banks of the valley follow the same law, being steep when concave in plan and less so when convex, so that the present course is but a minor variation on that which has been followed from the beginning. The winding of the valley, especially between Laifour and Revin, is so remarkable, as to suggest that circumstances at an early stage in its history were more favourable than now to deviation. For instance, at the latter town, where the Meuse forms a loop something like the impression of a horsehoof in mud, the place of the 'frog' is occupied by a low rounded hill, separated from the main upland by a neck of land, on which the town is built, standing not many feet above the level of the water. The current has pressed on the banks north and south of this headland ; the intervening crest, as the neck was narrowed, was gradually cut down by the action of rain, perhaps aided by melting snow, till at last it has almost disappeared. A similar loop may be observed at Fumay, where the river sweeps round another low

hill, against which it presses as it approaches the
little town[1].

The valley of the Rhine between Bingen and Bonn
affords a similar section (figs. 12, 13), except that the
flat parts, unused by the river, are perhaps more
frequent and certainly on a larger scale. The Moselle
for many miles above Coblentz has a general resemblance to the other river, but in this valley the tracts of
flat land by the water's edge are less conspicuous. But

Fig. 12.   Diagram of Rhine valley near Godesberg.

it is worth noting that in all these cases the valleys of
tributary streams which, as descending from the upland
have a more rapid fall, are generally V-shaped.   In
fact, the trough section, the height and slope of the
sides and the breadth of the base, depending on circumstances, are characteristic of a stream sufficiently
strong and full to overcome the obstacles of its
environment, but not enough to make it act as a saw.
Though nothing more than a brooklet, perhaps not

[1] A map of the former district is given in *Les Roches dites
plutoniennes de la Belgique et de l'Ardenne Française* (Poussin and
Renard), Pl. vii. and in *The Earth* (Reclus), p. 357.

even a drop of water may now be flowing along a valley which presents these outlines, they show that this was not always so, and that we need not hesitate to recognise it as a memorial of an old and fairly powerful drainage system.

In this connexion may be mentioned a singular, apparently anomalous, feature in some hilly limestone districts, namely the occurrence of valleys which are either always, or at least usually, quite waterless. These prove on examination to be, not deserted fragments of some earlier valley-system, but complete

Fig. 13.   Section of Rhine valley.

in themselves, exactly like those elsewhere worked out by streams. Such valleys are not uncommon in the chalk downs, especially to the south of the Thames valley. Green turf often covers bed and sides, and they resemble in general outline those down which water still runs. Not seldom, however, after descending one of these for some distance from its head, we come to a stream, taking its rise from one or more springs, which, in certain cases, only run for part of the year. Such intermittent streams generally go by the name of bournes, but they may be regarded

as a local accident rather than as calling for an
explanation altogether different from that of the
ordinary dry valley.   It is now generally agreed that
both the one and the other admit of the following
explanation.   The first beginning of the valley must
be carried back to a time when the mean temperature
was lower and the vegetation in the neighbourhood
much more scanty than it is at the present day.
Then the ground might be frozen, either permanently
or during a considerable part of the year, to a depth
of a few feet.   This would prevent the percolation
of water from rain and melting snow, and compel it
all to run off above ground, when it would produce
the usual effects.   Thus these valleys were the work
of rain and streams, but under conditions different
from those which now prevail.   In addition to this,
when the mean temperature became higher and this
icy crust no longer formed, the level of underground
saturation in a porous rock would be higher, so that,
if other circumstances were favourable, springs might
break out farther up the beds of the valleys than at
the present day.   But as the climate approached its
present normal, and the saturation level sank, the
springs, or the upper of them, if more than one set
existed, would cease to run.   Thus, in one place a
valley may now be entirely dry, in another it may
pass from this state after a wet season to one watered
in the usual way, while in a third, when the level of

the saturation surface temporarily rises above that of the valley floor, a spring may break out of the ground, and flow from one of the ancient outlets, forming a bourne.

In Hertfordshire there is a well-known bourne near Berkhamstead, which only runs after an exceptionally rainy season, usually beginning early in the spring and ceasing in the summer. According to Mr J. Hopkinson[1] it is recorded to have flowed twelve times in the interval between 1852 and 1904, and this has occurred when the rainfall for the year, ending on March 31, has not been less than thirty inches, and for half a century it has only twice failed to make its appearance, when this quantity has been measured.

But dry valleys with a bolder outline are rather common in districts where a limestone, such as that called the Carboniferous, attains a considerable thickness. We have only to study the noted Cheddar cliffs in Somersetshire to see that they have been carved out by a stream which has now disappeared (fig. 14). From the village of that name, at the foot of the Mendips, a glen runs up into the hill district. Though of no great extent, it is comparable "with some of the finest glens in Derbyshire. The cliffs rise on the one hand in absolute precipices, on the other they

---

[1] *Geology in the Field* (*Middlesex and Hertfordshire*), p. 11.

Fig. 14.   Cheddar Cliffs.

are, as a rule, less steep, but ascend very rapidly in
broken slopes and terraced walls of rock. The valley
bed is just wide enough, probably after some aid
from Art, to allow the carriage road to pass....The
rocks in the gorge are grey and bare on the right
bank, relieved by grassy slopes on the left. Here
and there perhaps a shrub or stunted yew has struck
its roots in some inaccessible cranny. Glossy ivy and
other climbing plants cling to the crags and redeem
the scene from absolute barrenness. The view in the
steepest part of the gorge is really grand. At one
point it takes a double curve beneath two lofty
pinnacled crags, which recede on either side from
a rocky bastion [1]." But though the glen bears every
feature of a waterworn valley, such as we might find
in Derbyshire (fig. 15) or among the limestone uplands
of Yorkshire, no stream runs down it. The Cheddar
caves, however, at a lower level, near the opening of
the valley are hardly less noted than the Cheddar
cliffs. Here, as in other parts of the Mendips
already mentioned, the hills are pierced by ramifying
caverns through some of which water still finds its
way, and no doubt the stream which carved out this
glen (and it is not the only one of its kind in the
Mendips) cut its way down to a level, where it found
that an underground passage afforded an easier way,

---

[1] The author, *Our own Country*, vol. IV. p. 30.

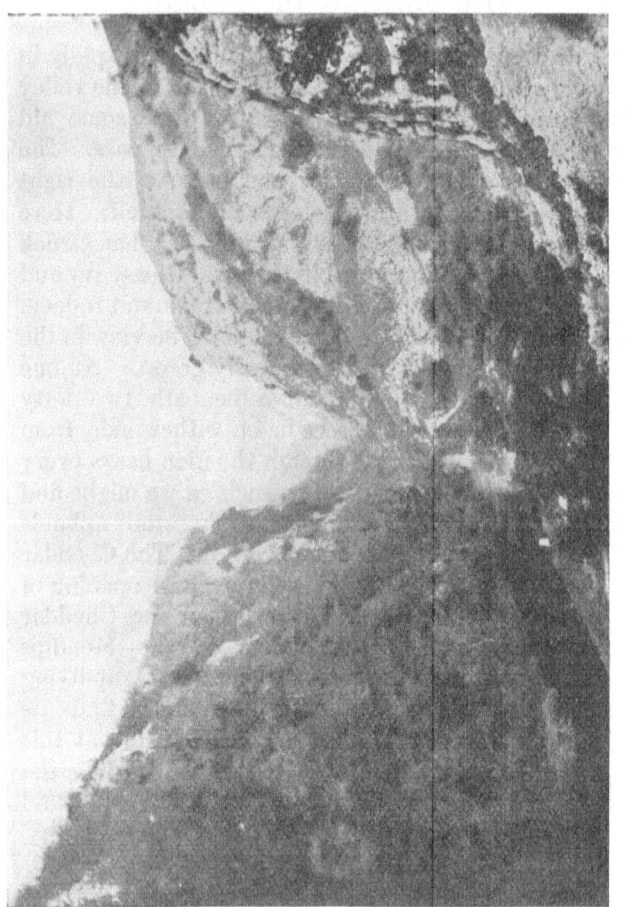

Fig. 15. Dovedale from Reynard's Cave.

and left off the sawing of a glen, in order to dig out a line of caves.

When rivers pass from a hilly to a lowland region, their courses widen, their banks become lower and their windings increase. They now flow over a very gentle slope ; their velocity diminishes, they gradually become incapable, under ordinary circumstances, of deepening their channel, and at last, as we shall presently see, deposit much of the material which they have been transporting. Their courses to these regions accord with two empiric laws, enunciated about half a century ago by the late James Fergusson[1]. These are (1) the extent of the curve is proportional to the volume of the stream ; and (2) the radius of curvature is proportional to the slope. When the fall is ten feet in a mile, the curve is almost a straight line : when it is three inches, that becomes nearly a semicircle. With such a curve the chord is about four times the width of the stream. With a fall between six inches and one foot the stream oscillates once in six times its width ; when it is more than a foot the oscillation is once in ten or twelve times the width. In other words big rivers, when they flow slowly, make big curves, and little rivers, little curves. Proofs of the latter rule are common in our own country, wherever a stream

[1] *Quart. Journ. Geol. Soc.* xix. (1863), p. 322.

remains intact. A brook may be seen wriggling like
a snake through the meadows of a flat-bottomed
valley, and as its volume increases, so does the magnitude of its curves. The Thames, for instance, upstream above Battersea makes one long sweep round
Fulham, another, in the opposite direction, round
Barnes Common, a third round Chiswick, yet another,
slightly more complex, round Kew and Richmond,
and behaves in the same way higher up from Twick-
enham.

The quicker the flow of a stream the straighter
its course and the greater its downward as compared
with its sideward action. A differential elevation of
the land, which increases with the distance from the
coast, or a greater rainfall in the same direction,
will produce a quicker flow and diminish the winding of a channel, while a change in the opposite
direction will have a contrary effect. The flat bed
of an English river-valley sometimes preserves, at a
slightly higher level than the present winding stream,
a fragment of a straighter course. The alteration
might be caused by a change of level, but is more
probably due to the fact that when the average temperature was lower than now, melting snow and a
heavier rainfall made the rivers larger and their
courses straighter. The strength of the excavating
tools, as a further study of valley structure will
demonstrate, has been liable to variation.

Nowhere is this more clear than in the cañon district of Colorado, which has been so admirably illustrated and described by the United States Geological Survey. A vast plateau extends from near the base of the Rocky Mountains to the head of the Gulf of California. Through this the river has cut a channel, nearly 500 miles in length, the bed of which is from 3000 to 6000 feet below the surface level. The plateau consists of a thick group of sedimentary rocks, resting on a solid floor of granite, which are of various ages, Palaeozoic, later Mesozoic and earlier Kainozoic. In the grandest part the section is chiefly cut through strata of Carboniferous and Permian age. The bedding is practically horizontal, and harder alternate with softer deposits. Two distinct phases are manifest in the history of the valley. The upper part, perhaps 1000 feet deep, is a broad trench with steep walls of alternating crag and slope, their edges occasionally notched by short ravines. Near to these insulated masses rises here and there from the bed of the trench, like outlying forts, remnants of once continuous material, which has been swept away. In the floor of this trench another valley has been cut, narrower, deeper and even steeper, but in its main characteristics similar to the other and older one. Cliffs, some 3000 feet in height, descend to the river, which generally washes their base on either side. The brink of these also is occasionally notched, or a

bastion tower may stand slightly in advance of the
main central wall.   One of these cañons would be an
ideal boundary for any country, since it makes a more
effective barrier than most mountain regions or a
narrow sea.   A hunter on one side might kill with
his rifle a deer on the other, and spend some days
before he could reach the venison.

Three or four favourable conditions must coincide
before one of these cañons can be formed : (1) The
existence of a plateau composed either of a thick
mass of a uniform hard rock, or of alternating strata,
which though differing somewhat in their power of
resisting denudation are not readily destroyed. (2) A
strong and rapid river making its way downwards
over the plateau.   (3) On this, however, the rainfall
must be small, for otherwise the cliffs would be too
much breached by the drainage from either side.   In
other words the cañon region must itself be arid, but
there must be a considerable rainfall on that from
which its rivers are supplied.   But the Colorado
cañon requires another condition : its first stage was
a valley, about six miles wide.   The stream, which excavated this, though powerful, must have flowed more
slowly than the present river, for, as we can hardly
suppose it to have occupied the whole breadth of the
valley, except perhaps during floods, it must to some
extent have wandered from side to side.   Then something happened to quicken its pace : probably the

whole region began to rise, and this not quite uniformly, but with a slight upward tilt inland. This
movement might also slightly affect the rainfall, increasing it on the mountainous zone and diminishing
it on the table-land[1]. In any case the cutting power
of the river was increased, and that, as we may infer,
from the sharp definition of the lower gorge, rather
suddenly. But after this had happened the rise
probably continued, slowly and steadily, so that the
conditions of erosion remained practically uniform
for a very long time[2].

The erosive action of running water is commonly
gradual and continuous, though of course it is always
intensified in time of flood (fig. 16). But it is impeded
by vegetation which, on the gentler slopes, may largely
neutralize the effect of rain : the blades of grass and
other herbage, like little shields, break the force of
the falling drops ; their roots bind the soil together
and save it from being washed away. But a single
storm in arid regions, sudden and violent, such as
sometimes occurs, may produce more effect than years

[1] This would require that the dominant winds should blow seawards as is the case on western coasts of continents in the tradewind
zone.

[2] For illustrations and particulars see T. C. Chamberlin and
R. D. Salisbury, *Geology, Processes and their Results* (1905), pp. 91–
95, A. Geikie, *Text-book of Geology* (1903), Frontispiece and pp.
504, 1382, and C. E. Dutton in *The Second Annual Reports of the U.S.
Geol. Survey*, p. 49.

of ordinary rain on the grass-clad slopes of our English lowlands. Very soon, every gully becomes

Fig. 16. Fresh stream-course produced by sudden fall of rain, Perthshire.

a plunging stream, the bed of a glen, often dry for weeks together, is swept by a rushing torrent,

hurrying everything before it. In the neighbourhood
of Berbera (Somaliland) a sandy plain separates a low
range of hills from the sea. From a depression in
these a shallow sandy nullah descends. Nearly 200
yards from its brink, where it was about 80 yards
wide and two feet deep, a camp had been pitched
early in April 1910. A sudden storm one night broke
upon the upland, the water came down like a 'bore,'
swept away and drowned eighteen camels with three
of their attendants, and 'a lot of small kit.' Between
an officer's bungalow and Berbera 'the road was cut
in three places, two of them about 4 feet deep and
broad, the third about 4 feet deep and 6 feet broad
—all in perhaps 500 yards[1].' The long winding
valleys in the mountain group of the Sinai pen-
insula are generally dry, for rain seldom falls, except
when some fierce storm bursts upon the mountains.
Then roaring torrents rush down towards the sea,
sweeping gravel and boulders before them. One of
these ravaged the Wády Soláf in 1867, " when an
Arab encampment was washed away and forty souls,
together with many camels, sheep, and other cattle,
perished in the waters. Mr Holland was in Sinai at
the time of the calamity and narrowly escaped losing
his life on the occasion. He describes the scene as
something terrible to witness : a boiling roaring
torrent filled the entire valley, carrying down huge

[1] Letter from Lieut. F. G. B. Wetherall, dated April 19, 1910.

boulders of rock as though they had been so many
pebbles, while whole families swept by, hurried to
destruction by the resistless course of the flood....
A single thunderstorm, with a heavy shower of rain,
will be sufficient to produce these dreadful effects
and to convert a dry and level valley into a roaring
river in a few short hours[1]."

Nearer to our own country, where such torrential
downfalls are almost without precedent, I have seen
considerable mischief done by a single storm. One
of unusual violence in the summer of 1875, as I
happened to be descending from the Col de Voza
into the valley of Chamonix, drove me to seek shelter
in a cottage. Before I left I heard a peculiar
dull grumbling sound, and inferred from shouting
which I heard, that there was something amiss. What
this was we discovered on coming to the edge of a
stream which flows through the hamlet. Usually it
runs at the bottom of a furrow, some ten yards deep
and rather more in width at the top, which has been
cut into a mass of dark slate. That was now occupied
almost to the brim, by a foaming roaring torrent,
which was something like pea-soup, only black, and
was sweeping along slabs of the rock. Presently we
arrived at the main road down the valley. Its
makers had thought the stream, under ordinary

[1] E. H. Palmer, *The Desert of the Exodus* (1871), pp. 22, 151,
212.

circumstances, too insignificant to build either a bridge or a culvert, but had allowed it to dribble across the road. Now this was completely flooded for a length of about twenty yards, and in the middle the main body of the torrent, some five or six yards in width and one in depth, was rushing with a fury that showed the communication to be completely cut. We were in the position of the proverbial rustic[1], but in this case the river did 'run down,' and it became possible after waiting some twenty minutes to splash through three or four inches of mud.

But I once saw in the Tyrol the result of a bigger 'cloud-burst.' In a fertile and rather broad part of the Zillerthal, as it is approaching its junction with the Inn valley, the lower slopes on the left bank consists of lead-coloured schists, which are rather easily destroyed. About a week before we passed a sudden and violent storm of rain had descended on this, and evidently had left its mark. But the worst mischief had been done by two streams which had flowed from it across the bed of the valley. That was covered for a considerable distance away from the main rush of the water some inches deep with a 'sticky' mud which nearer to their ordinary channels became more stony till at last it was full of blocks,

---

[1]   Rusticus expectat dum defluat amnis ; at ille
Labitur, et labetur in omne volubilis aevum.

HOR. *Ep.* i. 2. 42.

not a few of which resembled portmanteaux in size
and shape. The débris had swept across the road
(from which it had been dug away); meadows, corn-
fields, thickets had been buried; one wooden hut had
been lifted up and carried for about two hundred
yards; another had been invaded, and its floor
buried for some depth under mud. And all this
devastation was the result of a single storm.

Floods are especially destructive at the foot of
mountain ranges. In this the Pyrenees are generally
worse than the Alps owing to the absence of great
lakes which minimize the effect of the swollen torrents. In 1876, I went by railway along the base of
that chain, and in going from one mountain resort to
another, I passed not rarely the ruins of bridges
which had been swept away by a recent flood. This,
as I was told in Toulouse, had carried away two
suspension bridges, its great stone bridge alone
surviving the rush. But even in Britain a destructive
flood is not unknown. One of the most memorable
occurred in the valley of the Findhorn early in
August 1829. It was described by Sir Thomas Dick
Lauder[1], a summary of whose account is given in
Sir C. Lyell's *Principles of Geology*[2], and a fuller one
has recently been published in a very interesting

---

[1] *Account of the Great Floods in Morayshire.*
[2] Vol. I. p. 345 (ed. XI.).

illustrated history of that river[1]. On the morning of
August 2 a strange black cloud burst on the summits
of the mountain range from which the Findhorn takes
its rise. For three days and three nights the downpour
continued almost without intermission. Torrents of
water rushed down the hillsides, and on reaching the
lower ground swept away crops, ruined mills and dams
and acres of cultivated land. The river rose several
feet, in one place seventeen, above its usual level,
sweeping bridges before it, and spreading havoc everywhere. After the river has reached the foot of the
mountainous district, the scenery becomes comparatively open and it flows for a time along a strath. Then
it enters a narrow glen, cut through hard granite and
gneiss (called the Streens), the cliffs sometimes rising
almost vertically on either side, down which it foams,
leaps and swirls, till it comes to the Old Red Sandstone
and the valley again expands. The waters banked up
against the narrow entrance, detaching huge masses
of rock, and as they rushed through the Streens they
carried away cottages and crops, burying good land
beneath sand and gravel. In one or two places on
the lower land the river cut itself a new channel, in
others crags were undermined and landslips caused,
and the tributary burns took their share in the work
of devastation. Floods also, often almost as serious,
occurred in other counties of Scotland ; for instance, a

[1] George Bain, *The River Findhorn from Source to Sea* (1911).

five-arch bridge built of granite across the river Dee
at Ballater in Aberdeenshire was completely swept
away ; and the river Don forced 'a mass of four or
five hundred tons of stones, many of them two or three
hundred pounds weight, up an inclined plane rising
six feet in eight or ten yards, and left them in a heap
about three feet deep on a flat ground[1].'

With these examples of what can be done by rain
in excess and rivers in their fury we pass on to their
work as carriers, which indeed, together with that of
deposit, has already been indirectly illustrated.

CHAPTER III

THE TRANSPORT AND DEPOSIT OF MATERIALS

As in destroying, so in transporting, running water
acts both chemically and mechanically.   The former
process does not require a long account because it has
been implicitly described in referring to the underground action of streams, but a few additional facts are
needed to emphasize the importance of the work that
is being carried on invisibly before our eyes[2].

[1] Quoted in Lyell, *ut supra*, p. 345.

[2] Many valuable data bearing on the composition of spring- and
river-water and the transport of material, chemically and mechanically,
will be found in *The Data of Geochemistry* by F. W. Clarke.   Bulletin
491 *U.S. Geol. Survey* (1911), chapters iii. vi. and xii.

In any limestone district the water is always hard, that is contains carbonate of lime in solution. This material—the so-called fur—is deposited on the interior of kettles and boilers far more rapidly than when the water comes from or passes over sandstone or granite. For instance the water of the Scotch Dee, which drains a region almost without any limestone, contains to each 100,000 parts only 1·22 of carbonate of lime and no more than 1·90 of other mineral salts ; but the Rhine near Basel contains for the same amount 12·79 of carbonate of lime and 3·33 of other salts, while the numbers for the Thames near Ditton, since limestones crop out over a large part of its basin, are respectively, 16·84 and 10·37 ; calcium sulphate being 4·30 of the latter. Its waters accordingly carry in solution more than eight times as much mineral matter as the Dee.

It has been estimated, and this may serve to give a better idea of the quantity which is invisibly transferred, that about a thousand tons of carbonate of lime passes daily under Kingston bridge. This, if it were reconverted into chalk, would form a block measuring at its end 10 feet by 20 feet and 75 feet long, and would suffice, in a single year, to cover an area as large as Westminster Abbey with a layer nearly 89 feet thick[1].

[1] A ton of chalk is about 15 cubic feet ; and the area of the Abbey is 61,729 square feet. (Fergusson, *Handbook of Architecture*, p. 891.)

But large quantities of less soluble detritus are also transported. Mountain torrents hurry on quantities of sands and stones ; the latter varying according to their strength from pebbles to boulders ; and as their pace slackens the heavier members of their load are dropped, for the moving power of running water varies as the sixth power of its velocity ; hence, if this be doubled, the stream can move pebbles sixty-four times as large as before. As the bed of almost any Alpine torrent shows, the boulders are sometimes a yard or more in diameter, and if we stand by its side, when it is much swollen, we can hear them 'grumbling' as they are forced along its bed. During this process the angles are worn away from the fragments, large and small, and the irregular block is smoothed down into a pebble. The late Professor Daubrée devised an experiment which showed that a journey of more than fifteen and a half miles was needed to roll angular fragments of quartz or hard granite into pebbles, but that they were afterwards much more slowly reduced ; in fact that the smaller the pebble the farther it could travel without material loss of volume.

The annexed table, quoted in many text books, will give a rough idea of the transporting power of moving water :

Velocity in feet per second.	Material capable of transportation.
0·25 (300 yards an hour) ...	Soft clay
0·50	Fine sand
0·70	Coarse sand and pea-sized gravel.
1·00 (about 1200 yds. an hour)	Gravel (French-bean size)
2·25 (about 2700 yds. an hour)	Gravel an inch in diameter
4·00 (about 4800 yds. an hour)	Heavy shingle.

The meaning of these figures will be better appreciated by remembering that the first of them represents the ordinary movement in one of the sluggish rivers in an East-Anglian lowland ; that the average pace of the Rhine is rather under 1½ miles an hour, and this in the Danube, for several miles below Vienna, varies (with the quantity of water) from 2 to 3 miles a hour. It must also be remembered that these statements are only approximate, for the velocity of the water in the middle part of a stream is greater than it is at the sides, because here it is retarded by friction, as it is also at the bottom.

As the slope of a river bed diminishes, the coarser materials are gradually dropped, but the results of this belong rather to their constructive work, which must next be described.

When a turbid river, like one of those nourished by Alpine glaciers, flows into a lake, it forms about its entrance a grey cloud in the clear water, like a cumulus in the blue sky. That is due to the fine mud, which is kept in suspension by the moving water, but slowly settles down as this is brought almost to rest by the resistance of the larger mass. From a boat we can watch this cloud rolling slowly on beneath an increasing thickness of clear water, till at last it disappears from sight. Yet even then it continues to drift beneath the surface. Thus the finer mud, brought down by a river, is transported for a considerable distance from its mouth, but the coarser materials—sands and gravel—are deposited close to the shore. In this way a delta is formed and the land slowly gains upon the water. But when the process has begun, the river, when in flood, not only carries down detritus of greater coarseness and in larger quantity than under ordinary circumstances, but also overflows its banks and deposits this upon the level ground beyond them. The quickened rush of water scours the main channel and may even deepen it, but sometimes an opposite effect may be produced, because, when the current begins to slacken, it deposits on its bed the heavier part of the débris which it is transporting. Thus many rivers tend to build themselves out of their own beds, and this may even bring about changes of channel and additions to the thickness of the delta.

The same process is carried on when a river enters the sea, though as a rule, owing to the greater distance of highlands or mountains, the coarser detritus bears a less proportion to the finer, but still the total volume is likely to be greater than when the course of the river has been comparatively short. Egypt, as Herodotus observed nearly twenty-three centuries ago, is the gift of the Nile. Each inundation deposits a film of mud on the surface of the valley, augmenting, however slowly, the thickness of the alluvial deposit as well as its fertility. An excavation at Memphis, in the earlier part of the last century, showed that the base of the pedestal supporting a colossal statue of Rameses was buried beneath 9 feet 4 inches of Nile mud. This, according to the chronology adopted, had been 3211 years in accumulating, which gives an average rate of $3\frac{1}{2}$ inches in a century[1]. In 1885, borings at Kafr-es-Zayat and Tantah, about half way between Cairo and the Mediterranean, were carried down in the one case to a depth of 73 feet, in the other of 84 feet, without striking the floor of the old river channel[2], but here, of course, the lowest deposits reached must be thousands of years older than the pyramid of Khufru. The delta of the Nile, however, grows but slowly seawards. The reason of this, as has long been known, is that a strong west-ward current sweeps this part of the African coast, prevents mud from quickly settling, carries it away into deeper water, and distributes it over a wider area on the bed of the Mediterranean.

[1] *Phil. Trans.* 1855, Part I., Lyell, *Principles of Geology*, vol. I. p. 431.

[2] J. W. Judd, *Proc. Roy. Soc.*, No. 240 (1886), p. 213.

Estimates have been made of the quantity of detritus carried down by other great rivers; all proving the large amount that is unceasingly, though almost imperceptibly, in process of transference. The Ganges has been estimated to bring down annually enough material to form 42 pyramids, all equal in size to the largest at Gizeh[1] or to cover 172 square miles with a layer 1 foot thick. After it has joined the Brahmapootra the two discharge into the Bay of Bengal, year by year, detritus amounting to about forty thousand million cubic feet, which would raise the thickness of such a layer to over 8 feet. The Mississippi, however, transports more débris than the Ganges, for the estimated amount would suffice to cover 268 square miles with a layer 1 foot in depth, while the Hoangho carries a still larger quantity, for in its case the area would be 730 square miles. If we imagine the material built up into a solid rectangular block, the base of which covered a square mile, the height of this, in the case of the Ganges would be 172 feet, in that of the Mississippi 268 feet, and of the Hoangho 730 feet.

But by no means the whole of the detritus which passes any point on the river is taken down to the sea. Part of it is carried in suspension in the waters, the rest is pushed along the bottom by the current.

[1] The base of the Great Pyramid is about 253 yards and its height about 160 yards.

When this begins to flow more slowly, as it generally does in the lower part of the river basin, first the larger pebbles are left, then the smaller, then the sand and so on, though these may all be carried farther from their source when the river is in flood. But at that time, if not artificially restricted, it overflows the flatter parts of its valley, and by so doing deposits on them (as in the case of the Nile) a considerable portion of the sediment which it is transporting. The river also, by constantly dropping materials on its bed, constantly tends to build itself out of its channel, thus increasing the frequency of floods, though to some extent neutralizing this by raising the level of the land on either side.

A large part, however, of the finer detritus, as in the case of the Nile, is carried down to the sea, and then, if not swept away by currents it forms a delta. The rate at which this increases can, in some cases, be roughly estimated.

A nearly level alluvial plain separates the Lakes of Brienz and Thun, though obviously they must once have been a continuous sheet of water. This plain has been formed by the combined action of two torrents, the Lütschine, which is fed from some of the largest glaciers on the northern face of the Central Oberland, and the Lombach, which issues from the Habkern-thal on the opposite (or northern) side of the valley and about a mile west of the other opening. The delta of the Lütschine, which now, after some canalization, discharges itself into the Lake of Brienz, forces the river connecting the two lakes to flow (as at Interlaken) beneath the cliffs on the northern side of the valley, but presently it is encountered by the delta of the Lombach which, though a feebler agent, is able, now that the full force of the other is no longer felt, to push the Aar gradually on to the southern side. The volume of these two deltas must be very large, for their greatest thickness can hardly be less than 750 feet[1]. The Aar also makes a delta of its own at the head of the Lake of Brienz so that only its very finest mud, just enough to cause the well-known difference in tint between the waters of the two lakes, can reach the delta of the Lütschine. That of the Aar extends up to the Kirchet Gorge, a distance of about 9 miles from its present margin.

The Rhone also has reclaimed a large area from the Lake of Geneva. Its delta begins just below the 'narrow' at St Maurice, and extends for a distance of about 15 miles to Villeneuve, where the width is nearly 4 miles. But since Roman times quite half a league has been gained from the Lake, for Port Valais, which is now at least that distance inland, was then on the shore.

[1] The greatest depth of the Lake of Brienz is given as 856 feet; that of Thun as 712 feet

The Lake of Lucerne affords similar evidence. The Reuss has formed a delta at its head, which extends from Erstfeld to Flüelen, a distance of about $5\frac{1}{2}$ miles, but here the present rate of growth can be more exactly estimated, for the course of the Reuss was rectified about the middle of the last century, and in the next 27 years a delta was formed at the new place of influx, the volume of which was estimated as upwards of 141 million cubic feet, or an average yearly deposit of above 5 million cubic feet.

Deltas formed in the sea by important rivers are usually, as might be expected, on a grander scale. That of the Po projects into the Adriatic for some 12 miles beyond the ordinary shore line. In making this, the river, as is often the case when approaching the sea, divides into arms and has more than once changed its main channel. But little water now passes along the Po Vecchio, which was virtually abandoned in the twelfth century, or the Po Morto, its course prior to 1390. The delta has also advanced rapidly of late years; partly because the soil has been washed down from the mountain slopes in consequence of the reckless cutting down of forests, and perhaps even more because the restriction of the river by banks, in order to protect the cultivated land on either side, has indirectly helped it to raise the level of its bed. On the northern side the delta is merged with that of the Adige; indeed, this part

of the Italian coast, for a long distance, has been
reclaimed from the sea by the deposits of a series of
rivers. The marshy border of the mainland is being
slowly extended across the lagoon towards the islands
of Venice. But the delta of the Adige has occupied
the whole of this space, and united itself with the outer
line of sandy islands, of which the Lido is one. The
delta of the Po projects seaward in a long irregular
triangle, and a fair idea may now be formed of its
rate of increase. The town of Adria, at the beginning of our era a seaport which gave its name to the
Adriatic, is now 14 miles inland. The gain between
the years 1200 and 1600 of this era is estimated at
82 feet annually, and at 230 feet from the latter date
to 1804; this great increase, which is still maintained,
being due to the causes already mentioned.

Yet farther south Ravenna illustrates the same
process. Nineteen centuries ago this was a great
Roman seaport, with its harbour sheltered from the
Adriatic by a line of sandy islands, such as the Lido
farther north. Once old sand-bars, these at some time,
geologically not remote, have been elevated above
the sea. The original sea-port was four miles from
the open water. Then the harbour was slowly silted
up and a new one was established at Classis, beyond
Caesarea (the suburb erected by Augustus) which was
nearly three miles from the south-east corner of the
old city. That was a flourishing place in the fifth and

sixth centuries, but now both suburbs and harbours
have disappeared.   Only the stately church of St
Apollinare in Classe, built in 534 A.D., stands in
mournful solitude among the malarious marshes.
Though it is a longer time than Byron thought since
the 'Adrian wave flowed o'er' the site of the 'im-
memorial wood'—the noted Pineta—this in classic
days, must have been an island, one of the chain
already mentioned.   Even around Ravenna itself the
ground seems to have been raised, for just north-east
of the city the pavement surrounding the Mausoleum
of Theodoric, erected about 530 A.D., is now quite
three yards below the surface.   And yet this part of
the coast appears to have slowly subsided, though
that movement probably goes back to prehistoric
times.   A boring was made some years ago at Venice
which passed through 572 feet of alluvial material in
which, however, an old land surface was occasionally
pierced, and more rarely a bed containing marine
shells[1].

The delta of the Mississippi is yet more striking
in aspect and is on a much grander scale.   The end of
it projects from the continent like an arm, "pushing
out into the sea and spreading its branches over the

---

[1] The piercing of an occasional land surface in thick delta deposits
has occurred elsewhere, so possibly one of the indirect results of river
action may be the production of slight depressions in the crust of the
earth by locally loading it with heavy masses of sediment.

water like the fingers of a gigantic hand...These
narrow banks of mud, brought down into the open
sea by the fresh-water, present a striking spectacle.
In several places these banks are only a few yards
thick...the soil becomes perfectly spongy[1]," only a
species of reed can take root in them, and thus
give them a little coherence. Farther on they are
replaced by islets and shoals of mud which are
washed away and form again, wandering, so to speak,
between the river and the sea, at the will of the
winds and tide. The great river rolls out to sea
through its several mouths, discolouring the water
for many miles and for a still greater distance continuing its work in slowly raising the bed of the
Gulf of Mexico.

Winding rivers, however, as may be inferred from
what has been said above, have a tendency to
straighten their channels and correct their own
meanderings. Of this also the Mississippi affords
very striking examples. On its seaward way over the
wide alluvial plain—also a 'gift of the river'—to the
south of Vicksburg it describes great horseshoe curves,
which are often several miles in length. The ends of
these it gradually brings nearer and nearer till at last
the intervening neck of land is severed, and a new

---

[1] Reclus, *The Earth*, p. 405. The length of this part, which
perhaps bears an even closer resemblance to a tree with roots, is about
eighty miles.

channel opened. Through this the main stream flows, so that the loop becomes a backwater. Then silt accumulates at the junction of the two and as the bar thus formed approaches the surface, reeds and other plants begin to grow on it, the leaves of which act as filters and increase the deposit, till at last, with the help of other vegetation, the loop, now called a bayou, is converted into a long pool, supplied perhaps by some smaller streams which once flowed into the main channel of the Mississippi. The process of building and destroying, of altering the position of channels and all the minor features of the geography of the valley bed of a large river, must have been almost incessant in the days before man became civilized enough to contend with its vagaries and protect by embankments the land which he was desirous of reclaiming.

This tendency of a river to vary its course and to split into channels, so characteristic of the lower end of a delta, may be noticed on a small scale at the foot of an Alpine glacier, if it happens to terminate, as a few of them now do, on the margin of a stony plain. It is still more marked, as I am informed, in Spitzbergen where a tract of almost level ground sometimes intervenes between the sea and the edge of the glacier, and it must have been a common feature, before man began to struggle with nature, whenever a great river from a mountain region

debouched on a comparatively level lowland. Here
the sudden check to the velocity of its waters would
cause it to deposit the coarser gravels over the first
few miles of its journey along this lowland. These,
in times of flood, would be swept away from one
place and transferred to another; any chance obstacle
might divert a current from this to that position in
a broad valley plain, so that an old course might
readily be deserted and a new one followed. The
river might be parted, as at the end of its delta, into
branches which might afterwards be reunited, while,
if the general slope of the valley were slightly increased
by some earth-movements, this would quicken the rate
of the current and set it to work in digging out a
deeper channel in the gravels, which not long before
it had been depositing. Changes of this kind are
often illustrated in the larger valleys of the Alps.
The river, after descending the last of the steep rocky
slopes or steps, down which it hurries in the earlier
stage of its existence, flows along a valley, often fairly
wide, and in places so level as to be quite marshy.
Such, for instance, is the course of the Rhone from
a short distance above Brieg to the low limestone
barrier at St Maurice. But when the ice retired to
its present limits, after its last great advance during
the Glacial Epoch, it must have left either bare or
covered with a very moderate thickness of débris,
the floor of a rocky trench, not a few feet below the

present level of the valley, which it has subsequently
buried beneath sand and pebbles brought down from
the heights above. Every tributary stream from either
side of this valley has of course contributed its share,
which, however, as a rule is small compared with that
transported by the principal river; but in still lower
parts of the Rhone, and in other Alpine rivers, after
their emergence from the mountain region, properly
so-called, we often find that they have cut deep into
gravels which they must formerly have deposited, and
which may be seen to rise in terraces to a height of
sixty feet or more above the present level of the
water.

Alpine torrents, when their velocity is checked
by descending from some upland glen on to a more
level valley, are at once forced to deposit the coarser
materials which they have been transporting. Alluvial
fans, piles of gravel sloping steeply upwards, shaped
like half of a cone, are thus formed, over which the
water descends, not seldom through more than one
channel. In the Alps the fans, now in process of
deposit, are generally not on a great scale, but they
were larger at an earlier stage in their history, and
those beds of coarse gravel in the main valleys, over
which the Rhine, the Rhone, the Isère, the Durance
and the Po, and other rivers are now passing seawards and through which they have sometimes cut
their present channels, are records of the time when,

owing to climatic changes, they flowed with streams fuller and stronger than now. The huge masses of conglomerate called Nagelfluh, from which the Rigi, the Rossberg, and other great hills have been carved, must have been deposited in Miocene ages by the rivers of which those known to us are but the degenerate descendants, and the great gravel plain of the Crau, which borders the Rhone for many miles to the south of Avignon, is a memorial of a yet grander river in bygone ages. Even in our own country the coarse gravels which form terraces in the lower parts of the valleys at successive though not great heights above most of our lowland rivers, must have been deposited by streams more rapid than the Thames and the Test, or the sluggish Cam and Ouse.

But the work of deposit is chemical as well as mechanical, though this mode is less universal than the other, at any rate to a perceptible extent. The stalactites which hang like giant icicles or festoons of drapery from the walls of caverns in limestone districts, the stalagmite which gathers in mounds on their floors or sometimes covers them with a stony carpet, are deposited by the water, which descends as rain on the surface of the ground, takes up a burden during its passage through the rock and deposits it by evaporation on emergence. 'Dropping wells' and 'petrifying springs' illustrate the same process and their waters sometimes flow along the top of embankments which have been formed by themselves. Such have occasionally been carried across another stream, an arch being slowly extended from the one bank to the other. Not seldom they build up masses large enough to be quarried for masonry. Mosses, reeds, leaves, and other vegetation, facilitate the process, indeed some authorities believe that the work of deposit is at least greatly aided by the indirect action of microscopic algae. Rivers also, which have carved their courses in a limestone district, on reaching a favourable place, deposit some of the material they are transferring in solution. The streams which descend from such districts in the Apennines to the maritime plain of Central Italy, deposit large masses of calcareous tufa or travertine which is frequently quarried for building. The great blocks, which form the columns and other parts of those grand ruins at Paestum, are full of cylindrical holes, large enough sometimes to allow a finger to be inserted. These have been left by the reeds around which this fresh-water limestone was deposited. The stone, though rough looking, is very durable, for after having been exposed to the weather for more than twenty-three centuries it is still in good preservation, and its lack of beauty has no doubt protected it from the spoiler; for the more attractive material afterwards employed by the builders in the Roman city has been

largely carried away to serve more modern uses.
A much more ornamental variety of this kind of limestone supplied most of the building material 'for
ancient and modern Rome, for the Colosseum and for
St Peter's alike.' It is obtained from many quarries
on the Campagna near Tivoli[1] and can be seen at its
famous falls in process of formation. Everywhere in
the splash and drop of the water the travertine is
growing; it accumulates, like a frozen cascade, in
masses on the rocks and spreads in thick sheets on
the lowland at the foot of the hills. This variety is a
more handsome and compact stone than the other,
though still retaining a more or less irregularly
cellular structure, and in colour it is a pale cream
tint, so that at a distance it much resembles the
close-grained limestone of Cretaceous age, which in
many parts of Italy is a favourite building stone.

Similar instances of the chemically constructive
action of water, in the form of rain and rivers, may be
seen in the limestone regions of many other parts of the
world, especially where a warm climate is favourable
to evaporation, but the mode of formation is the
same. Chemical precipitation is almost restricted to
regions of calcareous rocks, for under ordinary circumstances the amount of silica and silicates obtained by
a stream which passes over crystalline rocks is too
small to allow of any precipitation. The work of

[1] The ancient Tibur, from which its name Travertine is derived.

geysers is hardly an exception to this rule, for though
these hot springs, with a more or less periodic explosive
action, build up cones and terraces of siliceous sinter,
the material has been obtained by the water, not
when it was flowing on the surface of the ground,
but by underground percolation, after it had reached
a zone where for some reason or other the temperature
was exceptionally high. In arid climates certain
soluble salts, which have been obtained from rocks,
crystalline as well as calcareous, such as anhydrite
or gypsum, rock-salt or potassium chloride, may be
precipitated. In that way extensive deposits of these
minerals have been formed during past ages and
the process still continues, where circumstances are
favourable. Above and below the surface the transference of material is in constant progress, and this,
as has been already mentioned, sometimes produces
secondary effects of considerable local importance,
and these again, by interference with the natural
drainage of a district, may initiate a new series of
consequences.

Though rain and river action is theoretically
competent to reduce a land surface to a uniform level,
it may be inferred from the reasons already given
that this result will but rarely happen. Plains, for
the most part, are likely to be produced by the sea,
either by the levelling work of its waves or by the
silting up of its bed. When a surface of this kind is

raised above the water by a fairly uniform terrestrial
upheaval, a new coastal plain forms a border to the
old land area, upon which the streams issuing from
the latter at once begin to work. If that were
originally an upland and these streams, or any of
them, had not yet cut down to their base-level[1], their
velocity would be increased along the older part of
their course but checked on coming to the lowland.
Here each one of them will widen more rapidly than it
will deepen its bed, will perhaps form great horseshoe
curves and be divided by islands ; in times of flood it
will deposit débris on the plain on either side and will
thus increase its elevation.   But in any case a new
drainage system will be initiated, in which the transverse valleys, as they may be called, are prolongations
of those already formed in the older plateau and a
number of longitudinal valleys are commenced. If the
upward movement continue for a long time, two
dissected plateaux will be formed, at different levels,
like two floors separated by a step. The dropping
down by a fault or faults of one part of an old
plateau produces similar results, as appears to have
happened in some parts of the United States.
Displacements of this kind will have the effect of
quickening the movement of streams in the upper
portion of the plateau, thus making the valleys more

[1] That is the level at which they are no longer able to deepen
their channels under the existing circumstances.

cañon-like, as well as to a certain extent checking the denudation of the more exposed parts of its surface. Some districts suggest that great changes have occurred in the processes of denudation, and it is often a difficult task to infer from the isolated remnants of an old plateau the circumstances which brought about their formation. In the so-called Saxon Switzerland the Elbe flows at the bottom of a deep trench cut in a plateau, the undulating surface of which is five or six hundred feet above the level of the river, but from this rise insulated masses of crags, like gigantic ruined castles, which are also formed of Quader Standstein[1] and apparently are the relics of an older plateau. How these were carved out or to what extent they have been subsequently diminished by subaerial denudation is not clearly understood. It is, however, fairly certain that something—perhaps a general upheaval of the whole area—happened to the Elbe, which compelled it to concentrate its energies upon a narrower area, and had for its result the present river trench and the undulations of the plateau in which this has been cut. But changes have occurred, even since this trench became well-defined, for the Elbe may be seen, for example near the well-known crags of the Bastei, flowing, on its convex side, at the foot of a wall of sandstone cliffs and

---

[1] A formation which is approximately contemporary with the chalk of England.

leaving, on the concave, a wide area of level ground between itself and the other wall of the trench. From this we may perhaps infer that the last change has been a reduction in the volume of water.

This feature—a change in the size of the trench excavated, a diminution in magnitude and an increase in the windings of the stream which it has made and still flows along it—can be observed in most of the principal river-valleys radiating from Central Europe. The Rhine, for instance, in cutting through the Taunus highland plateau between Bingen and Bonn

Fig. 17. Diagram of right bank of Rhine between Bingen and Coblenz.

has carved the more winding and narrower valley, so familiar to travellers, in the bed of a broader trough, and we may infer from its present course that it now meanders to a greater extent than formerly because, as in the case of the Elbe, level tracts of land intervene betwixt its concave margin and the slope on that side of the trough. We may also infer that the river, at no very remote time, as it is reckoned in geology, was doing more to fill up than to deepen its valley, because its present channel cuts for some little depth down into beds of stratified gravel which must have been deposited by its waters. Its minor tributaries, we may observe, show a marked inequality in their levels. By them V-shaped valleys sometimes have been carved down to the bed of the Rhine, but in other cases they end almost abruptly high up on the steep slopes between which it flows (fig. 17). In the Alps these would be termed 'hanging valleys' and be attributed by some persons to a glacier overdeepening the main valley, but here we have neither need nor right to invoke any such agency.

CHAPTER IV

THE HISTORY OF A RIVER SYSTEM

As we follow a river from its source in a mountain district to the commencement of its long wanderings over an alluvial plain or to its confluence with the sea, we find that if we allow for changes in the rate of descent or in the nature of the rocks traversed, the slope of its bed and the form of its valley are both altered. But in dealing with one of the larger rivers we must not forget that the work of denudation may have begun at a rather remote geological epoch and thus its results may have been modified by subsequent earth-movements. It is

therefore often very difficult to recover the earlier
chapters of its history, and a restoration of the
circumstances, under which its course began, may
not seldom be almost impossible. Some of these
complexities will now be considered, and as the
history of valley sculpture is generally written in
bolder characters and in plainer terms on a mountain
district, like the Apennines or the Alps than on a
hilly lowland like most parts of England, our first
examples shall be taken from these. Some minor
features in the formation of valleys have already been
mentioned, such as the effects of harder, alternating
with or intercalated in, softer rocks, of the slope of
such masses, and the like, as well as the tendency to
confluence and the consequent increase of tributaries.
But we must now point out some other effects, which
are often not less important. When two streams
start on their journey from opposite sides of a
dome-like mass, such as was mentioned in an earlier
chapter, each one, as we have said, cuts into this not
only downwards and sideways, but also backwards.
Thus their heads will be separated at first by a
narrow neck of the original surface. But these heads
in receding, will ultimately meet, and after that has
occurred, the 'col' or water-parting between the
two streams will be lowered. This process may be
continued until the initial slope of the valley is
considerably reduced, which of course will affect the

rate of denudation farther down and cause the place
where that process gives way to deposition to move
gradually up stream. The lowering process may be
carried so far that the watershed almost disappears,
becoming at last merely a gentle swell of ground, the
highest point of which is not easily determined.

But this often is accomplished more effectively and
to a more striking extent by the head of one valley
receding more rapidly than that of the other. In
so doing it cuts back through the watershed and
gradually carves away ground which formerly belonged
to the other. By this act of trespass, it not only
affects the contour of the watershed, as will immediately be described, but also captures some
streams which were formerly tributary to the other
river. This will happen in the case of transverse
valleys, when the slope is steeper or the rainfall
greater on one side than it is on the other, and the
same may occur, though the circumstances are likely
to be less favourable and the process less conspicuous,
when the valleys are longitudinal. The Alps afford
many examples, and it will be better to describe one
or two of the more striking before quoting any from
our own islands. None is more notable than the
Maloja Pass, between the head-waters of the Inn and
the Maira. At the present time the Mortaratsch and
Roseg glaciers are the real birth-place of the former
river, for the Flazbach Torrent, which transports the

drainage from the northern face of the Bernina group,
brings down the larger quantity of water, but the
actual valley of the Inn continues from above Sa-
maden as a well-marked and comparatively broad
trough up to the Maloja Pass. In it a step rather
more than 100 feet in height and about two miles
from that village, ascends to the Lake of St Moritz.
From this the trough, with little change in form,
leads past the Lakes of Campfer, Silvaplana and Sils
(which may once have been a single sheet of water),
and continues for some ten miles to the pass. So gentle
is the rise in the bed of this trough that the watershed
is only about 100 feet higher than the St Moritz Lake.
But within a very short distance of the head of the
Silser See the traveller is confronted with a sudden,
almost a startling change. The undulations of ice-
worn rock cease abruptly, and he stands at the edge
of great cliffs descending to the Val Bregaglia, into
which the carriage road, swinging away to the left,
descends in a series of zigzags. It is a drop of fully
800 feet, but of the explanation there can be no
doubt. The Alps slope, as is usual, more steeply
on their southern than on their northern side, and in
addition to this the rainfall on the former is heavier
than on the latter. For both reasons, the erosive
action of the streams on the one flank is more active
than on the other. The original watershed between
the valleys of the Inn and the Maira must once have

lain some miles to the south, and probably some hundreds of feet higher, for this pass—only 5960 feet above sea-level—is the lowest gap in the Alps between Switzerland and Italy. But as the Maira deepened its bed it cut back into the ridge between itself and its less active neighbour, removed the landmark, and annexed territory on the other side. By studying the geography we can find approximately the original position of the watershed. From the Maloja Pass we look into the opening of the Forno valley, just beyond it and on the left bank of the Val Bregaglia, to which it also descends steeply. Still farther down is the opening of a second valley, the Albigna. But each of these, though their streams are now tributary to the Maira, makes an acute angle with the direction of its flow, as if it were part of the river system of the Inn, while the valleys farther down the Val Bregaglia take the usual course, like branches joining the main stem of a tree. The Val Marozzo, also, which joins the right bank of the Val Bregaglia and may perhaps claim to be the chief source of the Maira, corresponds in direction with the two just named. Thus in some past geological epoch, perhaps even so far back as the beginning of the Pliocene, the watershed between the great drainage channels, now represented by the Inn and the Maira, may have risen somewhere between the sites of Vicosoprano and Bondo, or at least ten miles south of the Maloja

Pass.   But at the latter village a valley from the
Pizzo del Ferro (one of the peaks at the western end
of the *massif* giving rise to the Albigna and Forno
glaciers) joins the Val Bregaglia at the usual acute
angle with its course.

The same physical features can be seen though
not so conspicuously on most of the great Alpine
road-passes. The steeper part of the ascent, especially
when going towards Italy, terminates before the
watershed is reached, the approach to which is
comparatively level. On the other side, the descent
is usually, at any rate for some distance, the more
rapid. For instance, the upper part of the Lukmanier
Pass is almost level for more than a couple of miles,
though here it is less easy to determine whether the
tributary to the Rhine or to the Ticino has been
the trespasser. The St Gotthard, the Bernina, the
Brenner, and the Genèvre Passes are comparatively
level at the top, and this is true, so far as I remember,
of all the important road passes, as well as of a
considerable number of small paths.

Longitudinal valleys also have trespassed in like
way. The division in the Pusterthal between the
waters flowing on the one side to the Adige and the
other to the Drave is most difficult to detect, so level
is the great trough between the Central Tyrol and the
Dolomite ranges. This trough on the western side
drops rather slowly down, the descent being at first

not much more than 100 feet a mile, but on the eastern side the infant Drave soon carves for itself a furrow and the last relics of the trough are speedily obliterated. Except as a case of trespass on the part of the latter river, the present structure of the Pusterthal is inexplicable, and the small trench which joins it near Toblach, or almost on the present watershed, must indicate an ancient tributary from a transverse valley in the Dolomites. This second but narrower trench runs with a slightly higher gradient for about six miles and then drops steeply down to the glen, which soon enlarges into the valley of the Piave. Fragments of the old watershed of the Dolomite range remain in the Cristallo and the Croda Rossa, but a breach must have been made in it at this position by the enlargement of a glen which has its beginning some distance farther to the west. In fact cases of trespass are to be found almost everywhere in the Alps, and sometimes on a much greater scale, though here the proof is more indirect. The curious break between Monte Rosa and the Strahlhorn at the head of the Gorner and Findelen glaciers is due to the enlargement and regression of the Val Anzasca, by which the ridge, formerly connecting these two peaks, has been cut away, though it may have risen a thousand feet or more above the present watershed[1].

---

[1] For a discussion of these questions see the author, *Alpine Journal*, vol. XIV. p. 221.

But the same story of the making and trespass of valleys is told, though on a much smaller scale and in less obvious characters, among the lowland scenery of our own country. No better illustration can be found than in the Kent and Sussex Weald. The chalk range of the North Downs, after sloping rather gradually upwards from the valley of the Thames ends in a steep escarpment above a flat valley. This range, after running westward for some distance, sweeps round towards the south, and is then continued eastward in the South Downs till it is cut short at the sea near Beachy Head. The valley, which it overlooks, is excavated in the Gault and follows a similar course. Within this a second range rises, that of the Lower Greensand, which follows a similar course though the southern extension of it is a less conspicuous feature in the scenery, and within that is a second and broader valley, excavated in the Weald Clay, a deposit very much thicker than the Gault. Lastly, as the central core of the district, comes the hilly mass carved out of the harder sandstones which underlie the Clay. The origin of these successive ridges and valleys, horseshoe shaped, with that group of hills in the place of the 'frog,' was the subject of much dispute till the middle of the last century. It was recognised, at an earlier date, that this was a denuded anticlinal, the axis of which extended across the Channel, though much of its eastern part

had been cut away by the sea, but many geologists
maintained that the same agent of destruction had
removed the softer deposits from the remaining
portion, so that the waves had washed the base of
the steep cliffs in which the Downs descend towards
the Gault valley, and had applied a similar treatment
to the inner ring of the Weald Clay. Very probably
the sea did plane away the summit of the dome, quite
early in the Tertiary era, because on the higher part
of the North Downs the basement bed (Oldhaven) of
the London Clay rests upon the Chalk, but it is parted
from that rock in the vicinity of the Thames by about
150 feet of the Lower London Tertiaries. Nor do we
know exactly at what period the sea last invaded the
district. Some think it may have returned for a time
in the Glacial Epoch, but this many would deny. It
certainly covered the chalk hills, now fully six hundred
feet above its level, about Lenham and in the north-
east of Kent, during the earlier part of the Pliocene
period, but at that time the great open valleys cannot
yet have been excavated. For tracing their history
with a masterly hand we are indebted to Messrs
C. Le Neve Foster and W. Topley. In their classic
paper[1] they reasoned especially from the drainage
basin of the Medway. Its upper waters are formed
by streams from minor valleys, longitudinal and
transverse, in the Hastings Sands district which effect

[1] *Quart. Journ. Geol. Soc.*, vol. XXI. (1865), p. 474.

a junction in the Weald Clay lowland. Then the
river after following a rather curved course through
the Lower Greensand, bends definitely to the north,
crosses the narrow Gault lowland, cuts right through
the chalk range of the North Downs and flows into
the estuary of the Thames at Chatham. In its valley
gravels are found at various levels to a height of
about 300 feet above the sea, which evidently were
deposited by the same river in past stages of its
history. One of these, at Aylesford, only about 40
feet above it, has yielded palaeolithic implements
with bones of the mammoth and extinct species of
a rhinoceros and horse. The higher gravels must
obviously be much more ancient than this one, and
the channel of the Medway, when they were deposited,
must have been not much less than 300 feet above
its present one. In other words that river, since
it formed those gravels, must have carved out a
valley quite 250 feet deep and 7 miles broad. That
the process was gradual is shown by the successive
beds of gravel on its flanks, and the conclusions
formed from the Medway hold good of the other
valleys in the Weald district. If, then, rain and
rivers can be proved to have accomplished so large
a share of the work, we may safely credit them with
the remainder, though here the evidence is less well
preserved, and the mass removed may have been
almost double the thickness of the lower part and

its area far more extensive, for the physical features of the district suggest a continuous history for the whole (fig. 18).

As the original slope of the Weald dome must have been rather gentle, the system of transverse

Fig. 18. View from the Devil's Dyke, Sussex.

and longitudinal valleys is not quite so precisely defined as in the Alps, but the larger rivers keep to the former in passing through the uplands of the Lower Greensand and the Chalk and more or less follow the latter on the softer Weald Clay and Gault. On the northern side of the dome, five rivers now

flow into the Thames or its estuary. At the western
end is the Wey which receives but a small supply
from the Hastings Sands, its chief sources being on
the Weald Clay ; it is followed by the Mole, the Dart,
the Medway, and the Stour, all rising among the
former rocks. On the southern side, reckoning in
the same direction, we have the Arun, the Adur, the
Ouse and the Cuckmere, after which the sea has cut
through the chalk of the South Downs into the rocks
beneath it and amputated the lower parts of the
valleys.

It is difficult to fix the precise date when rain and
rivers began their work in sculpturing the hill and
valley system of the Weald. It must have been
after the time when the Coralline Crag began to be
deposited, perhaps not till rather late in the Pliocene
period. Much depends on the age assigned to certain
gravels on the North Downs, at elevations of from
six to seven hundred feet above sea-level. In them
are some fragments from sandstones occurring in the
hills to the south, and as these could not have been
transported by rivers across the present valleys, we
must conclude either that the latter had not yet been
excavated, or that a temporary submergence of the
district and a considerable lowering of temperature
had allowed the fragments to be carried by floating ice.
Each hypothesis has its difficulties, but though no
doubt the valleys of the Weald, like those in other

parts of the English lowlands, have been enlarged
and deepened since the latest part of the Ice Age,
that which involves the removal of all the material
lying below an approximate contour-line of 600 feet
demands unusually active denudation, and we cannot
forget that a well-marked system of hills and valleys
had been excavated in corresponding parts of England
prior to, or at any rate very early in, that age.

But whatever be the date, the work of sculpture
must have begun, probably on the surface of a
planed-off dome, so soon as this appeared above the
level of the waves. Since then the movements on
the whole have been upwards, but there may have
been, indeed most likely have been, one or two in the
contrary direction. Also they may not have been
uniform, or in other words the axis of the dome,
instead of being a continuous curve, may have been a
more or less wavy line. Undulations in the surface
would affect the initial directions of the drainage
channels or in some cases might alter the relative
levels in one that was already formed. As these
minor complications could only be inferred from a
large scale map, and in some cases the effects of later
denudation may have entirely effaced those of an
earlier one, it must suffice to say that the two
troughs between the hill ranges, and especially the
broader one marking the outcrop of the Weald Clay,
exhibit a complicated, though inconspicuous, system

of rivulets and brooks, gradually combining to form
a set of small rivers, which have a general corre-
spondence in direction with the strike of the strata,
and flow into the main transverse channels.   Their
heads, as they rise on low ground, are seldom very
clearly marked, and they are connected by very shallow
'cols' or saddle-shaped depressions.   Such a feature
indeed is a common one in many parts of our English
lowlands and admits of more than one explanation.
Sometimes it may be the result of a trespass : the
feeders of one stream having cut back into the
territory of the other ; sometimes the watershed may
have been gradually lowered by the action of rain as
the two streams worked back at about equal rates in
the soft rock.

Any attempt to give a detailed account of the
physiography of the British Islands would be out of
place in these pages, if only because it would often
introduce difficult controversial questions, such as
the existence and action of ice-sheets, or the amount
and extent of submergence, but we may call attention
to two or three valleys which, while illustrating the
action of rain and rivers, present some features
of their own which are not easily explained.

The Dee and the Severn rise and remain for
some time no very great distance apart in the Welsh
mountains.   On escaping from these they flow over
a lowland, formed mainly of comparatively soft

Triassic rocks, and it is remarkable that the Severn emerges in a valley which points towards the north-east more distinctly than the one belonging to the Dee. But instead of following, and ultimately joining, the latter river, it creeps eastward along a great projecting shoulder of the Palaeozoic rocks, as if it were endeavouring to find an exit southward among them, bends resolutely in that direction near Shrews-bury, and at last forces its way through the narrow portals of Coalbrook Dale, after which it winds on its way over a rapidly broadening valley to the Bristol Channel. We may reasonably conclude those portions of the valleys of the Dee and the Severn, which lie among the older Palaeozoic rocks, to have been 'blocked out' at an early date, during, if not before, the close of that era. The surrounding highlands probably contributed to the materials of the Trias and may have been above the sea during more than the Jurassic Period, but while the chalk was being deposited they must have been reduced to hilly islands if not entirely submerged. When that sea retired, their valleys must have been almost choked up with the littoral or other materials which it had de-posited. These the rivers would clear out and would make their way over the gradually widening lowland between the mountains and the sea. Across this, both Dee and Severn, if the elevation had been uniform, should have taken a course to the east or south-east, instead of bending in the one case towards the north, in the other towards the south. But probably it was not uniform. For in the south-eastern part of England, there was some folding along east and west lines, as already mentioned, which may have affected a wider area, and this certainly was repeated on a larger scale and perhaps had an influence on much of central and eastern England towards the end of the Eocene Period. By one of these movements the two great Welsh rivers may have been diverted into their present paths and made tributary to that far grander river which once flowed along the valley now occupied by the Irish Sea. These movements also may have determined the singular course of the Trent, which, rising on the western side of the Pennine range, sweeps round its southern end and bends again to the north till it at last combines with the Ouse to form the Humber. This lower portion of its course may possibly have been determined at a very late geological epoch, for some authorities consider the conspicuous trough through the Lincolnshire Wolds, now followed by the Witham, to be an ancient channel of the Trent. The same movements may have determined the course of the Thames, which rises on the eastern slope of the Cotswolds, takes a transverse course through the uplands of the Middle Oolite and the Chalk, receiving important tributaries from broad longitudinal valleys in the Oxford and the Kimeridge Clays, and then glides on to its estuary over the softer Tertiary deposits.

One branch of a river system is sometimes tapped by the feeder of another stream which captures the water previously contributed to the former. That usually occurs when a longitudinal valley of the second system works back into a transverse valley of the first one, so as to cut it slightly below the level of its bed and thus divert its water through the new opening. When this has happened, the channel below the gap becomes dry, and must so remain until enough water can be collected from its flanks and bed to form a new stream. But all the other streams, including the captive, go on deepening their beds, so that ultimately the deserted channel forms an isolated and perhaps comparatively wide elevated trench among the hills, the ends of which may be slightly notched by runlets of later date. Instances of capture are comparatively common and truncated channels far from rare. The elevated moorland of the Cleveland Hills affords excellent examples, which may be studied with advantage on both sides of the railway from Pickering to Grosmont on the Esk. Here we find, at heights of about 600 feet above sea-level, sundry well-marked channels winding over the moors, which in section are broad but shallow kneading troughs, for they have rather steeply sloping sides and wide flat beds that are often either boggy or traversed

by some feeble runlet which at last begins to notch
for itself a channel. These channels are occasionally
dissected by a newer and very different set of valleys,
which quickly take the form of a **V**-shaped glen,
broadening and deepening as it descends towards
the sea. But according to the ordinary principles
of interpretation, the one set must have been the
channels of fair-sized rivers flowing with a moderately
strong stream[1], probably at a time when the district
was less elevated than now—models on a reduced
scale of the Rhine or the Meuse on their passage
through the Taunus plateau,—and the other set must
have been made by later streams as they cut back
into the escarpment on their hurried passage from
the upland to the sea. The Whitby and Pickering
railway near Goathland runs along one of these
deserted trenches, the head of which probably once
lay farther to the north but has been removed and
the physiography of the district completely altered
by the aggressions of Wheeldale Beck and other

---

[1] This view, at the present time, would be disputed by many
geologists, probably by a majority of the younger, who assert that
these channels were made by the overflow from certain lakes, the
waters of which were held up by great ice-sheets advancing from the
north and east (*Quart. Jour. Geol. Soc.*, vol. LVIII. (1902), p. 471).
But as I have pointed out, and hope some day to demonstrate in
greater detail (*Presid. Address to Brit. Assoc.* 1910 p. 28), the form
of these valleys is irreconcilable with any such origin. They are the
dry beds of streams which had already flowed far, not of those near
the beginning of their course.

streams which combine to form the Murk Esk. Examination of a map shows that where aggressions of this kind have not been made, the sources of the southward running rivers, which we may presume to represent more nearly the original condition (allowing for some overdeepening of their lower portion), are generally to be found north of the centre of the moorland.

The great river-valleys, in the uplands of north-central Europe, afford evidence, so far as I have seen them, of some change in the volume and strength of the river by which they have been carved. The Rhine, already mentioned, may serve as an example. It presents the apparent anomaly, though this to some extent may be explained by its passage through a region of comparatively soft rocks, that the valley between Basel and Bingen is much wider than between the latter place and the Siebengebirge. In this part the trench-like valley, with which all travellers are familiar, appears to have been cut down into the bed of a wider one, which, however, is less definitely marked, because it has been exposed to denuding influences for a much longer time than the other. Whatever may have been the original history of the Rhine valley[1] there can be no doubt that this river,

---

[1] Sir A. Ramsay thought that the middle part of the Rhine valley had been initiated by one which had carried southward a river from the Taunus highlands and that differential movements, connected with

from an early period in the history of the Alps, has
carried off the drainage from a large part of the
northern face of that chain; but at first it may have
been a larger, though probably a slower flowing river
by which the upper and wider trench was excavated.
Then, with an increase of velocity, but perhaps some
diminution of volume, it took a rather more direct
course and carved out the lower trench in which it
still flows.   The tributaries, which it had received
during the earlier period, would still continue their
work of denudation and deepen their valleys, while
others would be formed to cut back into the old
escarpments.   Then a time came, and that, in a
geological sense, not very remote, when the cutting
power of the Rhine was again rather reduced, it
ceased to occupy quite the whole floor of the trench
and began to wander a little from its former channel,
continuing in some places to wear away its bank but
in others to leave a small level tract of dry land.   This
is the third stage in its history, by which the valley
of the Rhine has been brought into its present
condition (figs. 12 and 13, pages 54 and 55).

To the present day some geologists maintain
that glaciers have greatly deepened the valleys in

one phase in the history of the Alps, had caused a reversal of the
drainage and converted the middle valley into a lake, by the overflow
of which, at its northern end, the present gorge of the Rhine had been
excavated.

mountain regions from which they have now partly
or wholly disappeared. The extent of their work, in
the writer's opinion, has often been much exaggerated,
for all that he either has personally seen, or has
learnt from photographs, of districts which have
never been exposed to the action of ice in this form,
has convinced him that the dominant outlines in the
latter differ but little from those in the former, so
little as to be imperceptible at the distance of a very
few miles. To his eyes, " the general outlines of the
mountains about the Lake of Gennesaret and the
northern part of the Dead Sea recalled those around
the Lake of Annecy and on the south-eastern shore
of Leman. The sandstone crags, which rise here and
there like ruined castles from the lower plateau of the
Saxon Switzerland, resembled in outlines, though on
a smaller scale, some of the Dolomites in the southern
Tyrol[1]."

It must not, of course, be forgotten that heat
and cold, especially where the thermometer ranges
from distinctly below to considerably above the
freezing-point, are important agents in denudation.
Constant expansion and contraction fractures exposed
surfaces of rock, causing frequent falls of débris
and the ruinous condition of so many rocky peaks
when they are not protected by snow. This process
is at work in the Alps and to a less extent in

[1] *ut supra*, p. 10.

8—2

British mountain ranges, and it probably produced greater effects during the departure of the Ice Age. At the present day it is potent in the great arid regions, where the wind also takes some share in the work of denudation, but there is perhaps a tendency, at the present time, to overestimate its powers ; also we must not forget that when a snow-cap conceals the summit of a hill, or the ground beneath a sloping surface of soft rock is permanently frozen, each of these acts as a check to erosion.

Rivers are levellers, but there must be limits to their destructive powers. Allowing for differences in the nature of the rocks from which a valley is excavated, the slope of the latter is not a straight line but a curve, which is steeper in its upper part and gradually becomes flatter till it reaches a point at which the stream can no longer deepen its channel unless its velocity be increased. Below this it begins to deposit the material removed from its upper portion and to construct instead of destroying. It has then attained its base-level, as this phase is now called, and though, in times of flood, it may again take up the work from which it had desisted, this will be only for a short time, and the effect will soon be neutralized by the material which it has spread over the ground immediately below. The widening of a valley may also have its limits, for though rain running down a slope removes loose material, the surface of this, under ordinary circumstances, tends towards a curve of equilibrium, and the obstruction of the herbage, which springs up upon it, may at last bring denudation virtually to a standstill. In regard to this, the estimates of the average rate at which rivers lower their drainage areas[1] may be unintentionally a little misleading, because the work done is not equally distributed, and the form which the denuded surface ultimately assumes, instead of being a dead level, is one very gently undulated, on which we can still find traces of the portions prominent in earlier days. The removal of asperities, the bringing down of the high and the raising of the low, is the work of rain and rivers, but we may doubt whether their denudation will ever produce a dead level, whatever their deposition may do. For this end the sea is a more efficient agent; it is the plane in Nature's tool-chest; rain and rivers are the chisels, the gouges and the saws.

The sculpturing of a flat surface, which has been upraised without flexure, differs materially in form, as has been pointed out by several writers, from that

---

[1] The following estimates have been given of this average lowering [A. Geikie, *Text-book of Geology* (1905), p. 589]

The Po,	one foot in		729	years.
The Hoang Ho,	,,	,,	1464	,,
The Rhone,	,,	,,	1528	,,
The Mississippi,	,,	,,	6000	,,
The Danube,	,,	,,	6846	,,

impressed on one which has been steeply bent, but in both there have been distinct stages of development. These can be recognised, ' An old valley and a young one have different characteristics, and the one would no more be mistaken for the other by those who have learned to interpret them, than the face of an aged man would be mistaken for that of a child[1].'

The remark is true not only of a valley but also of the whole system of earth-sculpture, in which the former may be said to play the leading part : for the mountain after all is but the remnant left from a mass of upraised land after the excavation of the valleys. When this process has begun and slopes are steep a stream cuts deeply into them in its early days; the steeper they are, the stronger its efforts ; gravitation gives it the vigour of youth; but as time goes on the curve of its bed is brought more nearly to the line of repose and the stream passes into the less eventful period of mature life. In the earlier days of mountain sculpture the main ridges are sharp, serrate, and crowned with pinnacles; the peaks are pointed, the slopes are steep, the crags precipitous, the valleys furrows, on a large or small scale. These features, indicative of mountain youth, can been seen in photographs (if we have not been

---

[1] A brief and clear discussion of these is given by T. C. Chamberlin and R. D. Salisbury, *Geology, Processes and their Results*, (1905), pp. 79–87.

there) of the Alps, the Caucasus, and the Himalayas; though we must allow for some difference, not only in the nature of the rocks, but perhaps also, where the chains are large and complex, in the antiquity of the component ranges. But as the ages run on, the heat and the frost more and more shatter each projecting tooth and chip huge fragments from the face of every crag. The mountain climber can watch the stones dashing down cliffs, and may be awakened in his bivouac among the peaks by the roar of rock avalanches. Every fragment on the long slopes of screes on the mountain side—it is well-illustrated by the 'Screes' at Wastwater—has fallen from above, has done its part in the work of bringing down that which is high. For a long season the work of destruction sharpens the ridges and accentuates the peaks, but a time comes when the very forces which have made them begin to produce a contrary effect; the crests are lowered, their slopes meet at a wider angle; the cliffs gradually lose their steepness; the screes beneath intrude farther into the valley, and their surfaces assume a flatter curve. Every feature of the scenery becomes smoother, rounder, more undulate in outline. The effect of age on the face of Nature, at any rate in the mountain, is the diminution of wrinkles and the removal of asperities; for which an analogy cannot always be found in the seven ages of man, notwithstanding the Justice in *As You Like It*.

But when a hill and valley system has reached old age, a change in its environment may stimulate it to a renewal of its youth. An increased rainfall or a change in the distribution, such as has been mentioned above, may cause its streams to flow with larger volume and greater velocity, or the latter effect may be produced by an addition to its elevation above sea-level.   Such a case has been already described, so that it may suffice to remark that more than one cycle of denudation can often be traced, though commonly the second begins before the first is completed.   In parts of England the lower slopes of valleys and hills are not seldom steeper than the upper, as if a more sharply accentuated type of denudation had been impressed upon one more gently moulded.   This is very conspicuous in the Alps, where, as already pointed out, the lower parts of the valleys, to a height of 800 feet or more above the present bed, are either V-shaped or rather steep sided trenches.

Neither must we forget that rivers do much in removing temporary obstructions. A fall of rock from a mountain side may block a valley and convert a river into a lake, or a mud avalanche down a lateral glen may produce the same effect, or even a moraine may sometimes act as a barrier[1].

[1] An instance of the first is the well-known Lago d'Alleghe under the Monte Civetta, formed by a fall from the Monte Pezzo in 1772, of

Obstruction of this kind is rare in our own country; British mountains are no longer on a scale sufficiently large to admit of these phenomena, but they are common in regions like the Alps. Falls of débris and consequent mud avalanches were frequent during the epoch of transition from the age of ice to that of history. Well-marked terraces are not infrequently seen cut, not from the live rock but out of débris, and at too high a level to have been reached by any flooding of the existing torrent. They are well marked on the sloping left bank of the Lütschine opposite Grindelwald, and even better on the same side of the Ticino between Airolo and the gorge of Stalvedro. The significance of some similar cases, though in these the débris has been mainly brought down the valley rather than derived from its flanks, has been recently well discussed by Professor Sollas[1]. It would be easy to multiply instances of the effects of rain and rivers, in their endless succession of *Geburt und Grab,* but these may suffice to show that, while earth-movements are essential to bring its crust within their influence, while the sea often is a most potent agent of denudation, as well as the ultimate recipient of the larger part of the material which they transport, and is also the source on which they

the second the temporary lake in the Vispthal which in 1905 sub-merged part of the railway from Visp to Stalden.

[1] *Ancient Hunters,* (1911), ch. i.

primarily depend (for without it there would be no
rain), while heat and cold, snow and ice, all play their
parts, the rivers most of all

> Draw down Aeonian hills and sow
> The dust of continents to be.

Fig. 19.   The Lune, Westmorland.   (A quiet stage.)

CHAPTER V

MAN'S LEARNING OF NATURE'S LESSON

ONLY in quite recent times has the work of rain
and rivers in sculpturing the face of the earth been
clearly recognised. Glimpses of truth were no doubt
obtained, now by one observer, now by another, at
a very early date. Herodotus, who, so far as we
know, was the first to travel for the sake of acquiring
knowledge, declared, some twenty-three and a half
centuries ago, that Egypt was the gift of the Nile,
and must thus have realized that the mud, which its
waters deposited year by year, had been transported
from districts nearer to its source, and could not have
been removed without altering their form. More than
a century later Aristotle endorsed his predecessor's
statement, and illustrated the tendency of rivers to
block up their mouths by the deposit of débris,
which their waters brought down, from the Black
Sea and the Bosporus where, sixty years before he
wrote, sundry ports had been accessible to vessels of
larger draft than they then were. About nineteen
centuries ago, Strabo, the first writer on systematic
geography, extended the idea of a gradual conversion
of an expanse of water, first into marsh and then
into dry land, by showing that other rivers which enter

the Mediterranean, form, like the Nile, extensive alluvial deposits at their mouths, and on low ground inland, and especially mentions as instances 'the plains of the Hermes, Caïster, Maeander and Caïcus as having been formed by the streams that flow through them. The deltas vary, he thinks, according to the nature of the regions drained, being most developed where the country is large and the surface rocks are soft, and where the rivers are fed by many torrents. He remarks that these accumulations are prevented from advancing farther outwards into the sea by the ebb and flow of the tides[1].'

Evidently Strabo had grasped, nineteen centuries ago, the true principle of denudation by running water, and even had perceived that the outflowing currents of the Bosporus and the Mediterranean Sea were caused by the escape of the surplus waters which had been discharged by rivers. But these guesses after truth by the men of olden time seem not to have made any lasting impression on their contemporaries or successors, perhaps because the scarcity of books tended to isolate observers and limit the spread of their results. If we may trust Ovid[2], Strabo's contemporary, the philosophic Pythagoras, five centuries before their age, incorrect as some of his cosmological speculations might be, had

[1] Sir A. Geikie, *The Founders of Geology* (1905), p. 29.
[2] *Metamorphoses*, xv. 262.

arrived at sound conclusions on certain elementary
principles of geology. 'I have seen,' he is reported
to have said, 'the dry ground become a strait, and
land replace the sea; marine shells lying far away
from the water and an anchor resting on high
mountains[1]. The speeding down of water has
changed a field into a valley, and the mountain is
washed down into the sea...the boggy ground is
converted into dry sands and the thirsty land changed
into stagnant marshes.' After this he describes how
rivers issue from, or are swallowed up by, the earth,
and how their underground passages can sometimes
be traced, together with other notes of physiographic
change, too numerous to be mentioned here, declaring
in the course of them, that there had been a time
when Etna did not exist and would be one for its
death.

The victory of Christianity stimulated deductive
rather than inductive reasoning and diverted thought-
ful men from the study of nature to that of theology.
With the fall of Rome and the decadence of Byzantium,
began a long era of turmoil and darkness, so that for
some centuries more was done for science by Moham-
medan than by Christian teachers. It was not, indeed,
till the spirit of the Renaissance caused a stirring in
the dry bones of scholastic philosophy, that men
began to enquire into the 'how' and the 'why' of

[1] Here perhaps the poet improves upon the philosopher

the world around them, and even then they were
mentally fettered, in the supposed interests of religion,
by a dogmatism which was both ignorant and tyran-
nical. At first, also, the significance of fossils
awakened more interest than the problems of earth-
sculpture, a result which is not surprising, because
their resemblance to living things more obviously
called for explanation. It was not till the sixteenth
century that men set themselves seriously to in-
vestigate this problem and the leaders, chief among
whom was that versatile genius, Leonardo da Vinci
(1452–1519), were not found in England. Indeed,
one of the first really to discuss questions of physical
geography and recognise 'the predominant influence
of running water in carving out the irregularities of the
land,' seems to have been Nicholas Steno (1631–1687),
a Dane by birth, a Florentine by choice. Geological
speculations were coming into favour, though they
were fettered, not by the facts of nature, but by the
supposed need of according with the Book of Genesis,
which, however, provided in the Noachian Deluge an
escape from some difficulties. In Britain the idea
that fossils were *lusus naturae*, notwithstanding all
that had been done to prove them to have formed parts
of living creatures, still found favour ; in fact with us
the progress of geology, whether physical or strati-
graphical, during, and for some time after, the middle
of the seventeenth century, was slow and erratic. In

France Jean Étienne Guettard[1], a native of Étampes, was the first to apply the principle, afterwards worked out with such signal success by British geologists, of interpreting the past by observing processes at work in the present. One memoir among his rather voluminous writings is devoted to describing the degradation of mountains by heavy rains, rivers, and the sea. As a boy he had played at the foot of a picturesque crag in the Fontainebleau sandstone, which had been worn by the action of the elements, into a rude resemblance of a woman carrying an infant. This, in the course of less than half a century, had crumbled away and rolled down to the foot of the declivity. In the same time other rocks had been exposed, and ravines cut into level surfaces, which had previously shown no trace of either the one or the other. He fully recognised the power of the sea as one agent of denudation, but proved that the task of rain and rivers was to sculpture the surfaces once raised above it, depositing one portion of the detritus thus obtained in its waters and the other on the more level parts of valleys.

Desmarest, De Saussure, and others continued Guettard's work on the Continent, while in Great Britain we were indebted to James Hutton and his friend John Playfair for placing the subject of earth-

[1] See Sir A. Geikie, *Founders of Geology*, ch. iv., for an interesting summary of his varied and great services to geology.

sculpture on a scientific basis. The former, in 1785, gave an outline of the results of more than twenty years' patient study, but the full exposition of his views did not appear till ten years later, and even then his book entitled *The Theory of the Earth, or an Investigation of the Laws observable in the Composition, Dissolution and Restoration of Land upon the Globe* was not completed when its author died, and its conclusions, had they not found an exponent in Playfair, might have failed to secure due attention. Hutton's teaching in regard to our branch of the subject, may be expressed in the words of his disciple: "When the usual form of a river is considered, the trunk divided into many branches, which rise at great distance from one another, and these again are subdivided into an infinity of smaller ramifications, it becomes strongly impressed upon the mind that all these channels have been cut by the waters themselves: that they have been slowly dug out by the washing and erosion of the land, and that it is by the repeated touches of the same instrument that this curious assemblage of lines has been engraved so deeply on the surface of the globe."

But notwithstanding the advocacy of these illustrious Scotchmen, the mists of error still lingered long in Europe and even in Britain. In the latter it was not till near the end of the first quarter of the nineteenth century, that the Huttonian theory found many champions. The leader among these was G. Poulett Scrope, a Cambridge graduate independent of a profession, who, in his book on Volcanoes, availed himself of an opportunity of laying down clearly the principles on which the earth's history must be written,—principles which, so far as this branch of the subject is concerned, were elaborated in his classic paper, on the gradual excavation of the valleys in which the Meuse, the Moselle, and some other rivers are flowing, read to the Geological Society in the beginning of 1830[1], thus extending the explanation which he had already given of the valleys in Auvergne. Charles Lyell, a member of the sister university, and even the more conservative Roderick Murchison, who had been converted by Sedgwick from a fox-hunting sportsman to a keen geologist, followed Scrope, though not without some hesitation. Lyell at the end of a tour with the former in Auvergne, during the summer of 1828, expressed his conviction that rain and rivers were the most potent, if not exclusive, agents in the excavation of valleys, and this view was expressed by the two friends in a paper read to the Geological Society at the end of that year. Yet Lyell published a paper in 1833 in which he attributed the denudation of the Weald to the action of the sea, and Murchison, even so late as 1851, maintained that the flint drift of that area was the

[1] *Proc. Geol. Soc.*, vol. I. p. 170.

result of a great number of rushes of water, which
had swept away the materials entering into the
composition of this drift together with the bones of
the mammals entombed in it. It was long before
geologists ceased to have recourse to deluges in some
form or other to float them out of difficulties, and the
term diluvium, useless and misleading as it is, still
lingers among continental workers, though fortunately
it has almost disappeared from scientific literature in
England.

But though a belief in the excavating power of
water was beginning to gain ground, the majority of
geologists still believed that terrestrial movements
were the dominant factors in the formation of valleys.
This opinion found a weighty advocate in the distinguished Cambridge Mathematician, William Hopkins,
in a paper 'On the Geological Structure of the
Wealden District' (read in 1841 but not published
till 1848). In this he showed that the sudden
upheaval of an elliptical area of the earth's crust
into a dome-like form would produce two sets of
fissures (the one concentric and corresponding in
outline with the exterior of that area, the other
radial) which were now represented by the longitudinal and transverse valleys of the district. The
mathematical part of the argument was, of course,
indisputable, and the strains to which such an area
would be subjected might cause a widening of

pre-existent joints and other disturbances which would
facilitate the work of water in the above-named
directions, but neither was evidence forthcoming in
favour of displacements of adequate magnitude, nor
does the present hill and valley system of the Weald
preserve a record of terrestrial disturbances.   Earth
movements, as we have already seen, are sometimes
factors of real, occasionally of primary, importance
in the making of valleys, but this explanation of the
Weald proved to be a scientific tragedy, like that
attributed by Huxley to Herbert Spencer, where a
beautiful theory was killed by an ugly little fact.

But rain and rivers before long found a sturdy
and pugnacious defender in Colonel George Green-
wood, who in 1857 published the first edition of his
work entitled *Rain and Rivers, or Hutton and
Playfair against Lyell and all comers*, in which he
dealt especially with the Weald.   The book was full
of acute observations and sound inductions in regard
to the main point, but it failed to produce the influence which it really deserved.   One cause can be
best expressed in the words of a reviewer of the
second edition[1]: "We scarcely know what to say of this
extraordinary book.   That it contains many happy
theoretical hits cannot be denied, and though the
author's propensity for humourous digressions con-
tinually disturbs the grave attention of the reader,

[1] *Geol. Mag.*, 1867, p. 412.

some portions are well reasoned and clearly expressed. One of his great objects seems to be to attack Sir Charles Lyell, Professor Sedgwick, and other great founders of the science of Geology ; and while he accuses Humboldt of concealing the laws of Nature 'behind the double veil of Greek and Latin,' he scarcely writes a page without introducing a Latin sentence or quotation." Another cause of the failure was that the author actually maintained that man may have existed during the Silurian period and that "myriads of species of megatheriums, dinotheriums, anoplotheriums or *anyothertheriums* (*sic*) may have existed before the Silurian or primary and metamorphic period, without a vestige of their fossil remains being found in these strata." It is not therefore surprising that this advocate of rain and river action sometimes did his cause more harm than good.

It was not till 1862 that the work of rivers in the excavation of valleys was placed beyond question in our own country. For this we are indebted to another Cambridge man, J. B. Jukes, who on June 18 of that year, read his memorable paper ' On the mode of Formation of some of the River Valleys in the South of Ireland[1].' That paper, to quote the words of Sir A. Geikie, 'started the vigorous study of the origin of landscape which has been so characteristic a feature in the geological activity in the last half century.'

[1] *Quart. Journ. Geol. Soc.*, vol. xviii. (1862), p. 378.

In a postscript to his paper Professor Jukes referred to the Weald in the following words : "My acquaintance with the Weald of Kent is too super-ficial to allow me to express an opinion ; but perhaps I may venture to ask the question whether the Chalk, when once bared by marine denudation, which perhaps removed it entirely from the centre of the district, has not been largely dissolved by atmospheric action ; and whether the lateral river-valleys that now escape through ravines traversing the ruined walls of Chalk that surround the Weald may not be the expression of the former river-valleys that began to run down the slopes of the chalk from the thin dominant ridge that appeared as dry land during or after the Eocene Period ? "

It would be difficult to find words more instinct with scientific prescience, yet tempered by scientific caution, and the hypothesis thus advanced had not long to wait for demonstration. That was given by the classic paper of C. Le Neve Foster and W. Topley to which we have already referred[1]. In the course of it they demonstrate that no evidence can be found by field-work to confirm Hopkins' hypothesis of fracture, and that the following among other reasons show the inadequacy of the one of marine denudation: (1) 'The foot of the Chalk escarpment, and also that of the Lower Greensand, are not at the same level

---

[1] See p. 103 and *Quart. Jour. Geol. Soc.*, vol. XXI. (1865), p. 443.

all round the Weald as every sea-cliff must necessarily
be. This inequality of level can hardly be explained
by unequal elevations during the last rise of the land,
as the lowest parts are at the river-gorges. This
would necessarily be the case, if these transverse
valleys were cut down by running water.' (2) 'The
escarpments follow only the strike of the beds,
changing their direction as the strike changes. The
British islands, from the number of formations
exposed, and their great extent of coast, should
furnish some examples of long lines of cliffs following
the outcrop of beds, if any ever occur. But we find,
on the contrary, that the sea cuts across all formations alike, quite independent of the strike.' (3) We
never find any accumulation of shingle or any other
marine deposit at the foot of the escarpment,' an
argument which, as they observe, had been used by
Sir R. Murchison, in the paper already mentioned,
against the idea of denudation by the waves of the
sea. (4) That, as Professor Ramsay had pointed out,
'if the Weald were now submerged so as to convert
the escarpments into cliffs, we should have an
arrangement of sea and land in which denudation
could act but very feebly. There would be a central
group of islands surrounded by a strip of water in
the Weald Clay valley ; then a long ridge of Greensand country ; beyond this a second strip of water
washing the foot of the Chalk escarpment. "This

form of ground," remarks Ramsay, " would certainly be peculiar, and ill-adapted for the beating of a powerful surf so as to produce *on one side* only the cliffy escarpment that forms the inner edge of the oval of chalk."'

Two years later these arguments were driven home in a paper by W. Whitaker[1] 'On subaerial Denudation, and on Cliffs and Escarpments of the Chalk and Lower Tertiary Beds.' This was read to the Geological Society on May 8, 1867, but instead of appearing in its *Journal,* was published in the *Geological Magazine* for that year[2], from which fact we may infer that influences, adverse to Huttonian views, were still strong on the Council of that Society. Its author brings forward a number of examples from the Cretaceous and Tertiary rocks in the south-eastern portion of England, to illustrate the contrasts between escarpments and sea-cliffs. These he sums up under the following heads: (1) Escarpments always run along the strike, cliffs rarely do so. (2) The bottom of an escarpment is not at one level throughout. That of a sea-cliff is. (3) At the foot of an escarpment one does not find a beach or other trace of the action of the sea, but often such débris as would be left

[1] It is to the credit of the British Geological Survey, that Ramsay, Jukes, Le Neve Foster, Topley and Whitaker were all among its members.

[2] pp. 447, 483.

by a slow and quiet denuding power. (4) Two escarpments, facing the same way, often run near and parallel to one another for many miles. Not so with cliffs. (5) The ridge of an escarpment is a nearly even line and forms the highest ground of the neighbourhood[1]. The top of a cliff is often very uneven, and bordered by higher ground.

But though the battle, little more than half a century ago, was not quite won, the great importance of meteorological agents, and especially of rain and rivers, in sculpturing the earth's crust had become clearly recognised. The 'fissure theory' of valleys lingered for a time. In an introduction to the Geology of the Alps contributed in 1863 to Ball's *Guide to the Western Alps* by an eminent foreign geologist, E. Desor of Neuchatel, valleys of disruption are mentioned as one of the three types among which the valleys of this chain may be distributed; the others being valleys of outcrop and valleys of depression. The first he says 'are evidently produced by rents that have torn asunder ranges once continuous,' and he gives as examples, 'the valley of the Rhone between Bex and Martigny, and that of the Arve between Cluses and Sallenches; the valley of the Hinter Rhein above Coire, including the famous defile

---

[1] This statement holds good in the country examined, but would perhaps require to be a little modified in applying it to a mountainous region.

of the Via Mala, and the middle part of the valley of the Salza.' Every one of these would now be accepted without hesitation as an ordinary case of fluviatile erosion, and the final blow to the fissure theory was dealt, though in an indirect manner, by A. Ramsay in his paper on the origin of the greater Alpine Lakes[1]. These also had been claimed as gaping fissures in the earth's crust, but he shows that we have only to draw a transverse section of any one of them on a true scale to see the impossibility of the notion, that section having a far closer resemblance to the curve of a saucer than to the walls of a crack. He attributed these basins to the erosive action of glaciers during the age when they extended down the Alpine valleys far beyond their present limits. How far this hypothesis was correct is still a matter of controversy, but after the publication of his paper little more was heard of the fissure theory of valley-making. It has indeed been resuscitated of late years in another form, and in appearance rather than in reality, for the term 'rift-valley' is becoming a favourite with some of our younger geologists. This, however, is not at all what was meant by a fissure-valley, for to a considerable extent it is simply a mistranslation, and not seldom a misapplication, of the well-known German term 'Graben-versenkung' or 'trough sinking,' known for long to English

[1] *Quart. Jour. Geol. Soc.*, vol. xviii. (1862), p. 185.

geologists as trough-faults. Such displacements, as we have already said, do occur. Fissures, indeed, have been opened by earthquakes as at Charleston and in Calabria, in Japan and New Zealand. Lyell mentions one formed on the Southern Island, in the neighbourhood of Cook Strait, which could be traced for at least 50 miles, and on the two sides of which there had been a displacement in the level of the ground of as much as nine feet[1]. Similar fissures opened at the time of the Tarawera eruption in 1886, but we must not forget that these are generally on a comparatively small scale and are soon modified by the action of rain. Important faults, and especially trough-faults, must, however, in some cases greatly affect the courses of streams. This may be in either of two ways. If the fault (or group of faults) runs across the course of a river already determined, it may completely divert the water supplied by the tributaries above the line of fracture into another channel and leave the part below it a comparatively dry valley, down the ample bed of which an almost insignificant stream wanders seawards. Such I believe to be the history of the plain of Esdraelon or Valley of the Kishon in Palestine[2]. It must have been carved out by a stream of far greater volume than that river, which was probably supplied by tributaries from the Syrian Highlands on the

[1] *Principles of Geology*, vol. ii. (1872), pp. 82–89.
[2] See the author, *Geol. Mag.* 1904, p. 575.

Eastern side of the Jordan. Two of these may be indicated by the singular gaps on either side of Jebel Duhy, though the latter are partly due to the trespass of streams now descending to that river, which had its course determined by the pair of parallel faults[1] that can be traced from the Lebanon to the Red Sea. Thus the Jordan Valley as a whole is an example of the effect of faulting, for no sooner had this occurred, than the drainage of a large tract of the Syrian Highlands was compelled to take a southward course. The valley is also remarkable from the fact that so large a portion of it, from close to the Lake of Huleh to well south of the Dead Sea, is below the ocean level. But, though trough-faulting was the primary cause of the Jordan Valley, all its superficial features are due to ordinary rain and fluviatile erosion. The geographer might travel from one end to the other without suspecting that terrestrial displacements had been more important factors in determining the outlines of the scenery in this than in any other highland district. The mountains which rise on either side of the Lake of Gennesaret or of the Dead Sea present no marked difference from those beside one of the lakes in the limestone borderland of the Alps. Nothing in them suggests a rift. The displacement

---

[1] In cases of trough-faulting, the displacement may be due to two groups of parallel faults rather than to a single pair.

even at the first was a fault rather than a rift,
in the proper sense of that word, and its existence
is now obvious only to the scientific traveller. The
use of the term may be defended where the fractured
surfaces are so recent, in a geological sense, as to
have been little modified by the action of rain and
streams, as in the Great Rift Valley described by Prof.
J. W. Gregory in Masailand ; though even here, so
far as I can ascertain, the displacement is due to
'trough' or 'step faulting,' and if we may trust photographs the term rift-valley is not generally appropriate to the Lake Region of Africa, though the
indirect effects of trough-faulting have been most
important.   In fact, as has been pointed out in
an earlier chapter, without earth-movements there
would be no land for the waves to plane away, or for
rain and rivers to carve into mountains and valleys.
The earth's surface is like a palimpsest; the older
characters inscribed by the movements of the rocks
and the action of the sea have been so nearly
obliterated by the record of rain and rivers, that
their story can only be deciphered by close and prolonged investigation; but even the passing traveller
finds little difficulty in reading those which have
been and still are being traced by the agency of
falling or running water.

INDEX